'1'

'1'

The Foundation, Prediction, Verification, and Mathematization of Pure Being

Orest Bedrij

Xlibris

First edition 2008
Second edition 2009
Revised third edition 2014

Library of Congress Control Number: 2008906582
ISBN: Hardcover   978-1-4363-5789-0
               Softcover   978-1-4363-5788-3
               eBook   978-1-4500-4546-9

Library of Congress Cataloging-in-Publication Data

Bedrij, Orest, 1933—
p. '1': The Foundation, Prediction, Verification, and Mathematization of Pure Being /Orest Bedrij.—Revised third edition cm.

Includes bibliographical references and index.
1. '1'—The, Absolute Infinity, nature. 2. Physical quantities—Actual infinities. 3. Foundations of physical science—The, nature, attributes. 4. Mathematization of Physics—The, Logarithmic Slide Rule for Physical Relationships. 5. Fundamental Physical Constants—The, Derivation. 6. Time, space and gravity—Nature.

1. Title

Published in the United States by Xlibris, a partner of Random House Ventures, LLC, a subsidiary of Random House Inc.

Printed and bound in the United States of America.
10 9 8 7 6 5 4 3 2 1

Rev. date: 08/13/2014

We are participators in bringing into being not only the near and here but the far away and long ago. We are in this sense, participators in bringing about something of the universe in the distant past, and if we have one explanation for what's happening in the distant past why should we need more?

—John Archibald Wheeler

Many don't agree with John Wheeler, but if he is right then we and presumably other conscious observers throughout the universe are the creators—or at least the minds that make the universe manifest.

—Martin Redfern

Without a meaningful, believable story that explains the world we actually live in, people have no idea how to think about the big picture. And without a big picture, we are very small people.

—Joel Primack and Nancy Abrams

The universes come and go but you will always be around.

—Orest Bedrij

CONTENTS

PREFACE

When searching for harmony in life one must never forget in the drama of existence we are ourselves both actors and spectators.

—Niels Henrik David Bohr, 1885-1962

What is possible versus impossible depends entirely on what Universe you're living in. Until you understand the Universe you're living in, you cannot know what is possible

—Joel Primack and Nancy Abrams

We characterize '**1**' as "that which is," pure being: the background-independent first principle, ultimate reality. And so, as waves in water, whatever phenomena, relative reality, exist is in '**1**' and nothing can be considered or conceived without '**1**'.

In Exodus 3:14 '**1**' is represented as "I AM WHO I AM," or "I AM THAT I AM." In Revelation 1:8 '**1**' is described as "I am the Alpha and the Omega . . . who is, and who was, and who is to come, the Almighty."

In Svetasvatara Upanishad '**1**' is expressed as "Thou art woman. Thou art man. Thou art the youth and the maiden too. Thou as the old man totters with a staff . . . Thou art the seasons and the seas. Having no beginning, thou dost abide with immanence, wherefrom all beings are born."

The Chandogya Upanishad portrays '**1**' as "I, indeed, am below. I am above. I am to the west. I am to the east. I am to the south. I am to the north. I, indeed, am this whole world."

The utility of a clay jar or a room depends upon the measure of the *empty space* within the jar or the room. We utilize the '**1**' (the pure and formless *ground of being*, the quantity of the empty space in physics, the number of zeros in mathematics) in our *prediction and verification* of the fundamental physical constants and the fundamental laws of nature.

The *foundation* of existence and of change is the name we give to the absolutely first beginning (the pure and formless *ground of being*, the background-independent *first principle* and the highest final cause to which nothing else can be prior) from which manifestation arise. Understanding '**1**' (pure consciousness beyond space and time), which has no self-interest, but which nevertheless acts to uphold the world order at its ultimate level is a highly technical inquiry and a most exacting human activity. To get a quick overview, for the integration of a scattered and immense body of fundamental physical phenomena into a more systematic order, please refer to Appendix I. The 1993 paper is from the *Dopovidy* of the National Academy of Sciences in Ukraine. It is titled: "Scale invariance, unifying principle, order and sequence of physical quantities and fundamental constants."

"Can we ever expect to understand existence?" writes John Archibald Wheeler, a most distinguished physicist in his brilliant *At Home in the Universe*, a combination of scientific knowledge and openness to truth thinking. "Clues we have, and work to do, to make headway on that issue. Surely someday, we can believe, we will grasp the central idea of it all so simple, so beautiful, so compelling that we will all say to each other, 'Oh, how could it have been otherwise! How could we all have been so blind so long!'" [1]

Over the course of the past three decades, I have had the privilege of meeting Professor Wheeler, of cherishing his profound wisdom, unusual intellectual power, encouragement, and of sharing some of my work with him. Like Niels Bohr and Albert Einstein, his two inspirations of openness to truth (he would sit under their portraits in his Hightstown home), Wheeler was not afraid to explore the most fundamental problems of modern science and to acknowledge the depths of our ignorance. He once stated, "The time left for me on earth is limited, and the creation question is so formidable that I can hardly hope to answer in the time left for me."

Concerning one of my papers on the verification of ultimate reality, '**1**', through the laws of physics and direct experience, which has been discussed under different names: the theory of everything, the Absolute frame of reference, the ultimate value, the unity of knowledge, and the final law of nature [2], Professor Wheeler wrote to me, "I was not bright enough to find a prediction in your paper, so I will have to wait for some new development. In the meantime, you may get some comfort from my motto, 'If you would learn, teach.'"

Regarding the Upanishads, which deal with the quality of our life and the meaning and the value of *knowing* who we are (akin to the Socratic "know thyself") and *becoming* conscious of that which we are, Wheeler continues in the same letter, "Niels Bohr told me he went to India with no drive greater than the wish to discuss them with Indian scholars. He came away, he said, with the judgment they had great questions, but no answers. Yet someday we will have the answers."

This book is my response to Professor Wheeler's encouragement and support: "I was not bright enough to find a prediction in your paper."

My dearest reader, when you read '**1**', you will be happy and delighted to appreciate how close John Wheeler, Albert Einstein, David Bohm, Edward Witten, and many others have been to the central idea of it all. The central idea behind Bohr, Wheeler, Einstein, Erwin Schrödinger, and David Bohm's understanding of the underpinning of physical existence is that physics at its foundation is a unified whole. At the breadth and depth of absolute unity, '**1**', "the boundary of a boundary is zero ['0']," stated Wheeler. "This central principle of algebraic topology, identity, triviality, tautology, though it is, is also the unifying theme of Maxwell electrodynamics, Einstein geometrodynamics, and almost every version of modern field theory. That one can get so much from so little, almost *everything from almost nothing, inspires hope that we will someday complete the mathematization of physics and derive everything from nothing* [emphasis added], all law from no law ['**1**']." [3]

The purpose of this work is to increase the limits of human experience—to advance physics, perhaps the most fundamental science, and humanity in the search for fundamental answers to the mystery of the foundations of physical science. We deal with John Wheeler's and Niels Bohr's *prediction* and *verification* challenge of the fundamental physical constants and the laws of physics *without* cutting edge experiments. We present the central idea of existence in *unity*, *constancy*, and *differentiation*. We reduce it to *order* on a firm foundation, where all the laws of physics are integrated into a single all-embracing mathematical representation of the *least* that exists, the empty space—the laws of physics *unchanging frame of reference*, '**1**', via the mathematization of physics.

Let us go over the obvious. To make any sort of measurements concerning universal truths about nature, one customarily requires a standard, a common yardstick with a uniform scale of measurement. To apply a helpful cartographic or mapmaking metaphor, we are integrating galleries of "antique maps" of numerous scales (theories, principles, the laws of physics, the fundamental physical constants, and the Absolute and relative infinities) into *one* modern global map.

In antique maps there was no current Greenwich meridian line (the zero reference frame of longitude that runs from the North Pole to the South Pole), or the equator 0° (an equidistant reference frame on Earth's surface from the North Pole and the South Pole), or a uniform scale relationship between maps. Similarly, in current physics there is no acknowledged *laws of physics unchanging frame of reference,* '**1**', equivalently, the zero "rest frame" for the laws of physics and the physical quantities (actual, or relative infinities). Nor is there an accepted "initial conditions" of physics or a "uniform scale of measurement" for the fundamental physical constants and the unified theory of everything—a single principle that describes everything in the universe from subatomic particles to clusters of galaxies.

To apply one more useful metaphor, we take the Roman system of counting without the zero (the concept of the "zero" did not exist in Europe until after 1000 AD) and transform it into the present system of computation via the zero "rest frame" that does not spin and is not expanding in cosmology or physics.

Although zero is implied in every number, zero had no place on the *scale of numbers* for the Greeks, the Romans, and the pre-zero West. Similarly, although the laws of physics *unchanging frame of reference*, '**1**', is implied and enjoys experimental support in the speed of light in vacuum measurement, the '**1**' is not on the *scale* of the physical quantities and the laws of physics measurement and prediction of modern physics. Employing Phillip Edwin Peebles's, Albert Einstein professor of science at Princeton University, characterization, "one might well ask what the universe is expanding into, or where the space opening up between the particles came from." [4]

Why is this necessary? Using common mathematical tools, in an everyday computation, you feed numbers into the computer, and out come new numbers. In physics, you feed the laws of physics into the language of nature, and out come new laws of physics. The computations in the language of nature do not articulate and communicate in the ordinary numbers or equations of our computers. They speak in unity, constancy, and differentiation of *differential equations.* Differential equations provide the unifying mechanism of *how* the unifying process takes place. Calculus is the tool that we need to pose and solve these differential equations.

Just as with calculus in mathematics, so with calculus in physics we need the pivotal zero "rest frame," '**1**', and we also need the actually or relatively infinite of the physical quantities. Zero and (relative) infinity, which come in different sizes, are applicable to '**1**' and to all physical quantities generated within the '**1**'. Both are essential to physics but are now not in the measurement and prediction ground rules that describe nature with physics. We see electrons appear and disappear spontaneously in empty space, and we also see infinities in *renormalizable* quantum theories. Here, the actual values of the masses and the strengths of the forces cannot be predicted from the theory but have to be chosen to fit the observation.

The mathematical theory behind the differential equations can be viewed as the *unifying principle of order and continuity* function, the core of our understanding, behind diverse phenomena. The Schrödinger equation in quantum mechanics (QM), Maxwell's equations in electromagnetism, the Cauchy-Riemann equations in complex analysis, Laplace's equation that defines harmonic functions, Einstein's field equation in general relativity, the geodesic equation, the Black-Scholes equation in finance, Hamilton's equations in classical mechanics, and so forth are examples of the unifying principle of order and continuity in differential equations.

In our great intellectual adventure, the telescope and the microscope have given us the ability to discover the large-scale *and* the small-scale structure of the universe that has never been observed before. In the search for nature's ultimate laws, the logarithmic slide rule of physical relationships (LSPR), which is presented in this work and which serves as the *differential unifying principle* of predictive power, order and continuity in physical equations, gives us the ability to predict, discover, measure, and verify the large-scale and the small-scale structure of the fundamental constants and the laws of physics. The Schrödinger equation describes how the quantum state (also called a wavefunction or state vector) of a physical system changes in time. The LSPR *describes how the laws of physics and the fundamental physical constants change in reference to* the Absolutely Infinite "rest frame," '**1**', and the relatively infinite physical quantities (some can be vibrating strings).

Over the course of four decades, it has been an enormous privilege to have the opportunity to research and publish several works that deal with different aspects of existence and in the unification of distinct phenomena via the laws of physics. To weave it into the preexisting tapestry of the nonlocal unifying processes and the meaning of the counting system and the computability fundamentals, where by changing the term of an algorithm (a rule of procedure) we can transmit information faster than the speed of light. [5, 6] At '**1**' information travels (can reach the whole universe) instantaneously. [7] In these works we were not destroying the many beautiful fruits and hard labors of past achievements, but rather we were integrating them, like the mapmakers, within the laws of physics *unchanging frame of reference*, '**1**'. Making use of Stephen W. Hawking's expression, '**1**' "breathes fire into the equations and makes a universe for them to describe."

Similar to many others in the history of our planet, we have verified that the foundation of nature is an unchanging unity and profound beauty, simplicity, and absolute order. [7] Also that energy, the nature of time, gravity, the fundamental physical constants, and particles are not continuous but quantized [8] through the *spin* dynamics in '**1**'. We have attempted to encapsulate common features and principles of physical cosmology, mathematics, pure awareness, and the quantity-to-quantity, or atom-to-atom, spacing [9, 10] that Wheeler describes as "the boundary of a boundary is zero" and we characterize with the '**1**'.

Furthermore, we have considered the nature and evolution of the intergalactic medium, the origin of this structure, and the fact that the nature of our awareness and our actions determines what we perceive and experience in the universe. [2, 4, 7] Like Bohr, Wheeler, Einstein, Schrödinger, Ramanujan, Hawking, Steven Weinberg, and many others, we also found ourselves in the land of God, religion, and philosophy. [11]

In addition, in an effort to unify physics and astrophysics, we discovered that the unchanging "rest frame" *prior* to the definition of any metric structure or physical properties, in the laws of physics, is articulated with a mathematical expression of '**1**'. [8] Moreover, we found that the concepts of potentiality, hidden dimensions, entanglement, the boundary of boundary, and quantum nonlocality mathematics are implicit in the '**1**' (i.e., $1 = E/m_e c^2$). Where '**1**' is an exponent of '0': the *natural zero* '**1**'$\to$0 ($1 = 10^0 = e^0 = x^0$) [8, 9] and its inverse, the empty space, '**1**', as well as the unmodulated "rest frame" in measurements of physics and astrophysics.

The '**1**', as was mentioned earlier, has been elegantly verified and characterized with measurements, without our conscious realization of its profound magnitude and significance, by exact measurements of the speed of light in vacuum. Also, by exact measurements of the effect of our motion relative to the rest frame defined by the radiation and by the mean redshifts of distant galaxies, by tables of wavelengths and intensities of the principal atomic spectrum lines emitted by atoms in various stages of ionization and in the spectrum of the cosmic background radiation (CBR). "The spectra of distant objects indicate local physics is much the same everywhere," states Peebles, "the radio, optical, and X-ray lines, and the continuum between them, have the right arrangement and sensible-looking shapes, consistent with the standard physics of gas and stars." [4, 10] In our work we show *verified measurements* of '**1**' via the one-and-the-many principle, [2, 9] the laws of physics, the fundamental physical constants, the natural spectrum of physical quantities, and the LSPR. [8, 11]

On the path to understanding the foundation and the unification of the most enduring mystery in physics, we become conscious that many present-day physical models and theories are not a seamless whole. In fact, on closer inspection of the nature of unity in physics support unpredictability and contradiction that actually speak against convergence and unification. Specifically, although unity is fundamental to nature, physical science, and truth, we could not produce a "unified" relation and description of unity or a simple computer perfect physics [7] that combines science with balanced and readable representation and that is capable of communicating information effectively and quickly. Rather, these models are like the maps of all scales that have traditionally been drawn and made by hand without measurement and one standardized scale. [8]

When we substituted the fundamental physical constants into the classical relationships (Newton's theory, Maxwell's electrodynamics, and so on) and quantum relationships (Planck's constant, Schrödinger's equation, and so forth) our computer program would invariably crash again and again. This was also true with the very core of QM and the Heisenberg uncertainty principle equations, employing the mathematical constant "2π," and with Einstein's general relativity equations with the fundamental physical constants, the nature of potential and actual infinity, and so forth. [7] We realized that the present approach to physics was definitely unpredictable, inconsistent, and inadequate. [12, 13, 14]

Roger Penrose, Rouse Ball professor of mathematics at the University of Oxford, one of the great minds with uncommon intellectual power of the twentieth century and an exhibition of mathematical brilliance, characterizes the present-day situation in physics as "the major piece of 'patchwork' in modern physical theory, namely the way in which the classical and quantum levels of description are stitched together—very unconvincingly, in my view . . . . Perhaps the physicist's ultimate goal of a completely unified picture is an indeed unattainable dream." [15] Nancy Cartwright of the London School of Economics and Political Science, Lee Smolin of the Perimeter Institute for Theoretical Physics, Steven Weinberg of the University of Texas, and Stephen Hawking of Cambridge University, as well as others, have similar views.

There is an ongoing search by the leading minds of the world to extend our knowledge of the laws of nature, the unification of physical and astrophysical phenomena, [16, 17, 18] nature's final laws, [12, 13, 19, 20] and for the fundamental principles responsible within those parameters, cosmology, and the physics of pure awareness. By pure awareness we mean the nature of pure consciousness and the integration of mind and the quantum thought into physics. [11, 15]

Through direct experience of pure awareness, testing, and verification, my interaction with Professors W. I. Fushchych, J. Wheeler plus countless other people, and indirectly through the writings of H. A. Lorentz, Peebles, Planck, Bohr, Einstein, Schrödinger, Ramanujan, Weinberg, Hawking, and Penrose, I was inspired to understand the *nature* of unity and some foundational concepts in the unified physics. Thus, through this work, we have integrated a formidable array of experimentally verified findings, as well as an extraordinary range of confirmable accomplishments in physics, mathematics, and the principles of physical cosmology.

When you study this work, you may possibly agree with Wheeler's characterization. That when we reach the central idea, we will wonder why this was not apparent long ago, especially when we grasp that the laws of physics *unchanging frame of reference*, '**1**', has been right in front of us all the time. Indeed, it has been in front of us in the red shift of distant galaxies, in the tables of wavelengths of the principal atomic spectrum lines, in the speed of light reference, and in the one-and-the-many principle. "Oh, how could it have been otherwise? How could we all have been so blind so long!"

In four papers, 1-4, we have integrated the theoretical as well as the most profound experimental evidence of the foundation of physics—equivalently, the *pre*-big bang scenario of the "initial conditions," '**1**'. We characterize the nature of unity as *the unchanging reference frame*, '**1**', and describe a method for precise and rigorous prediction, validation, and the mathematization of physics. We show that scale invariance and several fundamental concepts of physics are missing from the basic embodiment of QM, classical physics, Einstein's general relativity, and the Heisenberg uncertainty principle equations.

The standard Copenhagen interpretation and other parts of physics and cosmology make numerous assumptions that have led to many heated debates concerning their interpretation. In fact, the less of reliable data and experimental support the fiercer the debates. At first, I did not know which equations were correct in the algorithmic compressions of '**1**'. As there was no room for error and I did not want to climb the wrong mountain, or leave any card unplayed, I was looking for *a common language that would bring unity to the sciences* and that would make predictions about the very laws of physics and the fundamental constants themselves. I have determined the laws of physics and the fundamental constants by integrating the '**1**' with *five* algorithmically compressed high-precision constants in which I had the highest confidence. Through this approach, I have achieved high accuracy results previously realized only through enormous labor and expense of high precision measurements.

Our verified predictions via this method are the following: the natural unit of mass, the Planck constant, the Josephson constant, the von Klitzing constant, the Compton wavelength, the atomic unit of velocity, the atomic unit of energy, the conductance quantum, and the quantum of circulation. These constants have a higher precision by one order of magnitude than the "CODATA recommended values."

In addition, using these five constants, we have calculated the *natural spectrum of physical quantities* [8] and fifty *new* physical constants that the CODATA do not provide. Also, with these five constants we go beyond the standard model of elementary particles and predict (1) the *nature of time* and the *nature of space*, (2) the fundamental relation of matter with the Planck constant, (3) how to deal with the gravitational force while presenting five *new* gravitational relationships, and (4) the *order* of the laws of physics and the values of their quantum electrodynamics (QED) constants.

Finally, we suggest that the advancement into the very foundation of the universe can be attained by means of the following: (1) the unchanging intervals in '1' (i.e., the laws of physics and the fundamental physical constants) and (2) the LSPR. [8] Similar to the logarithmic slide rule that is used for the prediction of *mathematical relationship of numbers*, the LSPR is a rigorously exact instrument used for the prediction of *physical relationship of quantities*, laws of physics, and the fundamental physical constants.

In the last few years, much has been written about the success and failure of string theory. [16, 21, 22] This field is full of controversial traps. "Many adherents and critics of string theory," says Lee Smolin, founding member of the Perimeter Institute for Theoretical Physics, "are so confirmed in their views that it is difficult to have a cordial discussion on the issue, even among friends." [22] "String theory, in a real sense, is the story of space and time since Einstein," [23] writes Columbia University's Brian Greene, one of the world's leading string theorists. We maintain that if we apply simple physical principles and confirmable fundamental basics correctly to the string concepts we will get the correct answers. We believe that some of the predictions and experimental evidence of string theory can be realized by adapting the concepts of time, space, and '1' presented in these papers.

The five basic rudiments between the foundations of mathematics now and mathematics in antiquity are in (1) the discovery and the significance of the zero, '0'; (2) the, . . . –2, –1, 0, 1, 2, . . ., positional place value notation that establishes a scale that the concept of '0' made possible; (3) the decimal point; (4) the discovery of logarithms that facilitated very large-scale computations through the addition and subtraction of numbers; and (5) the logarithmic slide rule, calculators, and computers. With these five basic elements, what had once been too difficult to calculate is now almost trivial. We are applying these five elements to understand the foundations of physical science and to attain the computationally classifiable mathematization of physics.

What is the pre-big bang scenario of the "initial condition" in science and in life? Our answer: the '1'—the laws of physics *unchanging frame of reference*, equivalently, the dimensionless zero point, '1', and its inverse, the dimensionless empty space, '1'. Here mathematics and physics correspond as one with absolute precision. We equate the hidden variables underlying the '1' (latent "software" attributes of '1' that await expression by the act of measurement itself) to algorithmically compressed imaginary numbers that were at first described as numbers that are not real. When we study the ground rules of imaginary numbers, we discover that the mathematics turns out to be simpler, even though the formalism seems to be more elaborate at first.

What are the most basic information elements of physics? Just as numbers are among the most basic information elements of mathematics, so the physical quantities (the *natural numbers*, actual infinities) in terms of which the laws of physics are expressed are the most basic information elements of physics in '1'.

How do we derive physical quantities, i.e., how do we actualize potentialities and hidden dimensions of the *natural numbers*, in '1'? Actualization of the potentialities and hidden dimensions takes place through measurement of the "interval" from one '0' rest frame point to another '0' rest frame point in the empty space. Schrödinger's equation in QM is an example. It describes how the quantum state of a physical system changes in time. The Schrödinger equation can be transformed into the Feynman path integral or into the Heisenberg formalism. We can use other differential equations to describe cooling in thermodynamics (Newton's law), population dynamics (the Lotka-Volterra equation), fluid dynamics (the Navier-Stokes equations), radioactive decay in nuclear physics, space, gravity, and so forth.

Similar to steps, these "intervals" in empty space represent different quantities of time, space, gravity, etc. Thus, as with numbers, the measured distance from '0' to –1, or 0 to –2, or 0 to 1, or '0' to 2 in '1' establishes the size and scale of the fundamental physical constants. We classify computationally the size and scale of the fundamental physical constants. A 'real number' has a place on the real number line; a "measured physical quantity" has a place on the measured *natural spectrum of physical quantities*.

As with numbers, . . . –2, –1, 0, 1, 2, . . ., in the *positional place value notation of mathematics* that establishes the size and scale that the concept of '0' made possible, so with physical quantities in the *positional place value notation of physics*, we establish the size and scale via the natural spectrum of physical quantities that the concept of '**1**' made possible.

The topological 4-manifolds are not computationally classifiable (time and space are independent constants) and, therefore, noncomputable in this missing physics. In paper 1, the four-dimensional space-time is separated into six dimensions of space (two are dimensionless, and four are dimensional) and three dimensions of time. The dimensionless space, '**1**', is fixed, while the dimensional geometry of space is not fixed but evolves with time.

The decimal point in mathematics is used to indicate the place where values change from positive to negative powers of 10. That is, the decimal point separates the integer part of a number from its fractional parts. We equate the decimal point in mathematics to the '**1**' in physics, which is used to separate the large-scale physics (group quantity) from the small-scale physics (individual quantity). It includes classical physics as well as quantum physics.

The fundamental physical constants on the same side of '**1**' as the number 1 represent a *number of* physical quantities or relative infinities (+∞). A physical quantity is relatively or actually infinite if it is simply unbounded. Those located on the opposite side of '**1**' from the number 1 represent *one* physical quantity and –∞. Therefore, in the wave (one physical quantity)/ particle (number of physical quantities) duality, there is a wave associated with every particle. The inverse is also true; there is a particle associated with every wave. In paper 1, we demonstrate this via the '**1**' and the one-and-the-many principle. With this principle we generate the nature of time, the nature of space, and the nature of gravitational field strength via motion in '**1**'. We also use this principle as the frame of reference to indicate the place where the values of the fundamental physical constants change from positive to negative powers to construct the natural spectrum of physical quantities.

In papers 1, 2, and 3, we show how five high-precision fundamental physical constants, like the five fingers of our hand, can do the work of producing all the fundamental physical constants in physics. In reference [15] Penrose writes, "I believe that the normal view of physicists is that, if we really understood quantum properly, we could deduce classical physics from it." Penrose is right. In paper 2, equations (2.10a), (2.12a), (2.17a), and

(2.18a), you will see that to obtain quantum level constants or classical level constants, we can realize this principle in terms of the same simple physics.

How do we achieve the mathematization of physics? We combine three natural spectra of physical quantities to make the LSPR. Like in mathematics, so now with physics, many computations and experiments that had previously required a great deal of time, money, and effort are now almost inconsequential, and many experiments that were once impossible to complete are now within reach. This development, when applied to physics, will further accelerate our progress, advancement, and openness to truth.

Stephen Hawking, Lucasian professor of mathematics at Cambridge University, in the concluding passage of his popular work *A Brief History of Time*, makes this observation, "If we do discover a complete theory, it should in time be understandable in broad principle by everyone, not just a few scientists. Then we shall all, philosophers, scientists, and just ordinary people, be able to take the part in the discussion of why it is that we and the universe exist. If we find the answer to that, it would be the ultimate triumph of human reason—for then we would truly know the mind of God."

Different levels of experimental science, integration, and unification produced distinct ways in which phenomena could be incorporated under a common experimental framework of the reality of '1'. However, it is important to realize that given primitiveness of our experimental technology our work of unearthing the '1' is embarrassingly rudimentary and has barely began. The Sumerian and Babylonian science lacked the conceptual framework of natural laws; we lack the conceptual framework of the observer, which is at the very center of the foundation of physics, in the role, function, and meaning of '1'. This becomes very apparent through the LSPR and the '1' which Wheeler calls "law without law," or "all law from no law" [3]. Here, the '1', the mysterium magnum of existence takes on different frameworks, expressions, and manifestations, which one of the most famous verses in the Torah (Exodus 3:14) characterizes as: "I am that I am," or "I-shall-be that I-shall-be." Niels Bohr stated, "in life one must never forget in the drama of existence we are ourselves both actors and spectators."

Where do we go from here? Freeman Dyson, professor of physics at the Institute for Advanced Study in Princeton, in his rare treat *Infinite in All Directions*, gives us a hint: "Mind has waited for three billion years on this planet before composing its first string quartet. It may have to wait for another three billion years before it spreads all over the galaxy. I do not expect that it will have to wait so long."

Enjoy it!

References

[1] Wheeler J A (1994) *At Home in the Universe* (AIP Press, Woodbury, NY), p 310.

[2] Bedrij O (2000) Revelation and Verification of Ultimate Reality and Meaning through Direct Experience and the Laws of Physics. *Ultimate Reality and Meaning, University of Toronto Press,* 23:1.

[3] Wheeler J A (1994) *At Home in the Universe* (AIP Press, Woodbury, NY), pp 303, 302.

[4] Peebles P J E (1993) *Principles of Physical Cosmology* (Princeton University Press, Princeton, NJ), pp. 134-154.

[5] Bedrij O (1962) Carry-Select Adder *IRE Trans Electronic Computers,* vol EC-11, pp 340-346.

[6] Bedrij O (1963) *Selecting Adder,* US Patent 3100,835.

[7] Orest [Bedrij] (1978) *One* (Strawberry Hill Press, San Francisco, CA)

[8] Bedrij O (1993) Scale Invariance, Unifying Principle, Order and Sequence of Physical Quantities and Fundamental Constants. *Proc Inst Math Natl Acad Sci* Ukraine; also, (1994) Ukrainian Mathematical Journal, tr of the Proc Inst Math Natl Acad Sci Ukraine (Allerton Press, New York), pp 67-73.

[9] Bedrij O (2002) New Relationships and Measurements for Gravity Physics. Fourth Inter Conf: Symmetry in Nonlinear Math Phys. *Proc Inst Math Natl Acad Sci Ukraine* vol. 43, Part 2, pp. 589-601.

[10] Harrison G R ed (1969) *Massachusetts Institute of Technology Wavelength Tables* (The M I T Press, Massachusetts).

[11] Bedrij O (2005) *Celebrate Your Divinity* (Xlibris, Philadelphia, PA).

[12] Greenberger D M and Zeilinger A (1995) *Fundamental Problems in Quantum Theory: A Conference Held In Honor of Professor John A. Wheeler* (The New York Academy of Sciences, New York, NY)

[13] Elitzur A, Dolev S, Kolenda N, eds (2005) *Quo Vadis Quantum Mechanics?* (Springer, Berlin).

[14] Stroke H H, ed (1995) *The Physical Review: The First Hundred Years* (Amer Inst of Phys, Woodbury, NY).

[15] Penrose R with Shimony A, Cartwright N, and Hawking S (1997) *The Large, the Small and the Human Mind* (Cambridge University Press, Cambridge, UK), p 180.

[16] Penrose R (2004) *The Road to Reality: A complete guide to the Laws of the Universe* (Jonathan Cape, London).

[17]   Weinberg S (1992) *Dreams of a Final Theory* (Pantheon Books, New York).

[18]   Morrison M (2000) *Unifying Scientific Theories: Physical Concepts and Mathematical Structures* (Cambridge University Press, Cambridge UK)

[19]   Block M M (1986) *First Aspen Winter Physics Conference* (The New York Academy of Sciences, New York) vol 461.

[20]   Kim Y S, Zachary W W eds (1989) *Proceedings of the International Symposium on Spacetime Symmetries in Commemoration of the 50th Anniversary of Eugene Paul Wigner's Fundamental Paper on the Inhomogeneous Lorentz Group* (North-Holland, Amsterdam).

[21]   Woit P (2006) *Not Even Wrong: The Failure of String Theory and the Search for Unity in Physical Law* (Basic Books, NY).

[22]   Smolin L (2007) *The Trouble with Physics* (Houghton Mifflin Company, New York, NY), p. xviii.

[23]   Greene B (2003) *The Elegant Universe* (Vintage Books, New York, NY), p. 6.

ACKNOWLEDGMENTS
AND GRATITUDE

We bow down in awe and gratitude for the past. Without all
that came before us, none of us would be awakening now!

—Barbara Marx Hubbard

It is a great pleasure to thank the National Institute of Standards and Technology, the CODATA Task Group on Fundamental Constants, and the countless other institutions around the world that facilitate the gathering of evidence and high-precision measurements and that disseminate the fundamental constants that make a mathematical description of this difficult area of inquiry possible.

A heartfelt thanks and appreciation go to Richard Knouse for applying his rich background in mathematics, programming, and computer modeling, highlighting the relationships among the laws of physics, the fundamental physical constants of nature, and the foundation of physical science, '1'. Very special thanks and gratitude to A. Stecyk, O. Trylis, and G. R. Talbott for testing the spectrum of physical quantities and the logarithmic slide-rule of physical relationships. I am especially grateful to Professor O. Pilskalns for calculating and verifying the "2006 CODATA recommended values" in the equations in this work. In addition, special thanks and appreciation to A. I. Lenec for careful editing and suggestions. Also, a profound thanks and gratitude to J. P. G. Suson and C. Elle for thorough copyediting, indexing, reviewing, and editorial guidance.

As always, I am very indebted and thankful to my countless colleagues and collaborators; and their kindness, wisdom, and creativity is gratefully acknowledged. Professors W. I. Fushchych, P. Kotzer, J. A. Wheeler, O. S.

‘**1**’

Parasiuk, R. Andrushkiw, W. Tiller, and T. Horvath deserve special thanks and gratitude for their very valuable conversations, encouragement, and insights to my voyage through the labyrinth of explanations and effective mathematization of the language of nature.

1

'1': The Foundation, Prediction, and Unification of Physics

Light brings us the news of the Universe.

—William Henry Bragg,
1862-1942

The quantum hypothesis will eventually find its exact expression in certain equations which will be a more exact formulation of the law of causality.

—Max Karl Ludwig Planck,
1858-1947

Abstract

In paper 1 we present the following: (1) the background independent "initial conditions" of physics, that is, the foundation of physical science and prediction; (2) the merging of the microscopic and the macroscopic worlds via the '**1**' in the one-and-the-many principle; and (3) the nature of time and the nature of space. In addition, we provide the five algorithmically compressed high-precision constants that generate the natural spectrum of physical quantities with a 2π spin dynamics QED constant.

1 Introduction

"Predictions can be very difficult," stated Niels Bohr on a number of occasions. [1] To predict precise expressions of the laws of physics, in a "dimensional uniformity of physical equations" that are more exact than a number of the present formulations of physics and that have profound implications for physics in the future, several foundational problems had to be addressed and solved.

First, QM and Albert Einstein's general relativity frameworks, as one, are incompatible. We know that the quantum theory is the most successful theory of all time, yet we also found that the very core of QM and the Heisenberg uncertainty principle equations employing the *mathematical constant* "2π" does not fit the facts as to the nature of physical reality when applied to spin dynamics and atomic structure. [2]

Second, the nonexact and nonexperimentally validated formulation includes Einstein's general relativity equation. Einstein's formulations lack dimensional uniformity of physical equations. [3] These esteemed *intuitive* formulations also lack rigorous measurement support of the standard mode that R. Penrose and others have already presented. [4]

The foundational problems addressed in this work are as follows: (1) the pre-big bang scenario of the "initial conditions" and dynamical evolution of physics had to be precisely and rigorously characterized and verified; (2) the merging of the microscopic and the macroscopic physical worlds by way of the "unchanging intervals" in the coordinate system at rest needed to be validated with measurements; (3) the "five algorithmically compressed constants" with which we could calculate the *natural spectrum of physical quantities* (i.e., the qualitative ordering, hierarchy, symmetries, and the scale of over seventy fundamental physical constants) had to be determined; (4) dealing with Werner Heisenberg's 1927 indeterminacy relations, the cornerstone of quantum theory, had to be clarified; (5) the QED quantization constant (allowing the classical laws to transform *simply* in one-to-one correspondence to the laws of QM) had to be calculated; (6) the nature of time, the nature of space, gravitational field strength, and matter had to be decoded; (7) over fifty new fundamental physical constants, for which measurements didn't exist in QED, had to be calculated; (8) a hierarchical order among actual infinity of infinities had to be solved; (9) as the terascale of the natural spectrum of quantities stretches out to over 10^{100} power, a method to facilitate the *mathematization of physics* on a handheld LSPR had to be developed.

Einstein's "coordinate system at rest," equivalently the "rest frame" or the "empty space," '**1**', [1, 3] and the unchanging intervals describing unity, derivatives, and relative infinity of infinities in '**1**', is characterized by the fundamental constants and physical quantities, presented in papers 1-4 and equations (1.xx)-(4.xx), and suggests a new experimentally supported formalism for advancing the rigorous prediction, verification, and the mathematization of physics. We do not describe the path to fundamental results in terms of the current standard model of particle physics or fields of force or geometry or space and time or even string-based cosmology. Rather, we are describing and validating it with measurements of the '**1**'.

There are many ways one can characterize the depth and richness of the Absolutely Infinite '**1**'. We shall name a few. We can characterize the '**1**' as the vacuum "rest frame" of the speed of light. Also, '**1**' is the background independent "rest frame" defined by the spectrum of the cosmic background radiation or the redshifts-distance relation "rest frame" for galaxies. In our interpretation, '**1**' represents the *foundation of physical science* from which we suggest all observational evidence arises with simplicity and mathematical elegance at one end and dimensional uniformity of physical equations and precision at the other. We merge the microscopic with the large-scale structure of the universe by introducing intervals (limits between events: the fundamental physical constants and physical quantities) in the empty space. This is achieved by an algebraic method of characterizing the changeless foundation, without reference to anything external, that we suggest is the existence of everything that has happened, is happening, and ever will happen in our universe. The suggested representation is an extension of the prevailing interpretation of the conventional approaches that exist today. [5, 6, 7]

"What is physics?" Although this question is perennial, physics has been usually defined as the science of matter and physical energy—fundamental particles, their associated fields and interactions between them, and the elementary constituents of the physical world that are susceptible to experimental investigation and theoretical inquiry. With the advent of quantum physics, particularly the indeterminacy principle and the inherent statistical nature of measurement of the very small, these basic assumptions have been considered and modified. [8, 9, 10, 11] We suggest expanding the advance of progress and great triumphs of physics by bringing all phenomena within "one" fundamental principle, the foundation of physics, or the "unchangeable." Physics may then be defined as the *science of the foundation of nature or the unchangeable, '**1**', validated by measurements.*

The immutability of the fundamental laws of nature is the central postulate of all science. The reason that the scientific method works, or that the spectrum of light from galaxies billions of light years apart is validated by repeatable corroborations, is that the "stationary ground state," the "rest frame" of the empty space, is invariant. Thus in any physical science, we are dealing with the fine-grained hierarchy of '1'. The invariance of the empty space, which itself is not influenced by physical conditions and in which the universe is expanding into, makes physics possible with deep implications. Richard Feynman put it this way, "If the same circumstances don't always produce the same results, predictions are impossible and science will collapse." [12] In effect, science venerates and exalts the unchangeable. Our principal reason for restating these common fundamental concepts is to make clear that we are dealing with well known, experimentally supported predictive tenets and physical principles that have heretofore not been capitalized on for the purpose of making well-substantiated predictions subordinate to all-embracing initial conditions.

From the vantage point of basic science, we characterize the '1' as the state in which metric relations (matter, gravitational force, electromagnetic fields, strings, etc.) no longer play the role of the highest common factor and are transformed to the null vector, 0, by use of any annihilation operator. Hence, the '1' is where the wave function collapses and the normalization condition (the shift in context) takes place (see the "One-and-the-Many Principle" below and the "Orders of Infinity" in paper 4).

Also the '1', which one could use to describe the initial conditions from which a dynamical evolution of Lorentz covariant quantum theory or a given set of physical systems evolve, [13] is the *natural zero*: '1' [14] and the "carry" in computer mathematics. [15] The '1' is analogous with, but not identical to, L. de Broglie and D. Bohm's "hidden" variables of the foundations of QM. [16, 17] Additionally, '1' is W. Heisenberg's world of "potential," [18] H. Pagels' preexisting "cosmic code," [19] E. Witten's topological quantum field theory, [20] and D. Finkelstein's prequantum network. [21]

The '1', where physics is constant like "hardwired software" in a computer, is as well a mega verse of $10^{\pm\infty}$ empty space nonconceptual states or potentialities—hidden dimensions, similar but not identical to H. Everett's many-worlds interpretation [22] or G. Garriga and A. Vilenkin's *Many Worlds in One*. [23] The universes of empty space potentialities themselves, similar to the LSPR, are *algorithmically compressed* and within reach *at every point* of '1'. This suggests that in the future we may be able to consult intelligent life in distant galaxies and understand what happened before the universe started expanding. In subsequent papers, a more general formulation of physics in terms of '**1**', the mathematization of physics, validation, and greater precision are provided.

2 '1' = The Empty Space

In physics (i.e., Einstein's $1 = E/m_e c^2$), '**1**' needs no point of reference or speculation that it is the resting frame or the natural zero, '**1**'→0 ($1 = 10^0 = e^0 = x^0$). When removed from equations of physics, however, to differentiate the 1 of the positive integers, or the 1 of the set of all real or complex numbers, from the natural zero, we add single quotes (' ') in bold as the marks for the natural zero, '**1**'.

Thus, the Absolutely Infinite first principle, can be characterized as the algorithmically compressed *dimensionless point* (the natural zero, or singularity at the initial state of the universe, the period prior to inflation, or the physical quantity formation): '**1**'. And its inverse, the *dimensionless empty space*: '**1**'. Moreover, '**1**' is the in equilibrio starting point to which all explanations may be traced, and by means of which, other physical objects can be explained. Furthermore, '**1**' is also a renormalizability principle and the boundary for space-time and between the quantum and classical realms where the same fundamental essence is observed in different ways by each observer.

In consequence '**1**' represents the unmodulated a priori pure awareness, where the nature of pure consciousness causes collapse of the wave function; the quantum thinking dissolves, mind becomes no-mind and the most fundamental thing that exists. Noise degrades pure awareness and the high fidelity of harmony, simplicity, beauty, and information processing. Pure awareness is an interval between one quantum thought and the next—one fundamental physical constant or physical quantity and the other. It is a state of no thought, the reality of emptiness, and the great author of nature. According to Bohr, "The task of science is to expand our experience and to reduce it to order." I agree with Bohr.

As we are ourselves doers, the viewers (observers) and the viewed (observed), we in reality generate our own universe. John Wheeler is right when he states, "The universe is a self-excited circuit." And as the boundary between the observer and the observed is obliterated in '**1**', every point or group of points in '**1**' (i.e., physical quantities) has its own awareness of the cosmos.

Everything arises out of '**1**' and dissolves back into '**1**'. The '**1**' holds atoms and galaxies together, drives nuclear and cosmic acceleration/deceleration, and determines the dynamics of atoms, the material universe, and patterns that govern the development of forms and change. If all the large-scale structure of the material world around us in the universe were smothered out into a consistent ocean of atoms, there would only be roughly one atom in every cubic meter of space. Indeed, very little. And the atoms are also empty.

The '1' is based on mathematical formalism automatically covering any procedure of measurement. This includes QM of individual electrons, atoms, and molecules and also the experimental apparatus and the physicists who use it. We know the '1' through our experience of it. Similar to retrieving information from computer software, *what we customarily describe in the laws of physics is the '1' exposed to our system of questioning.*

Nearly all models take for granted that gravity is the dominant force in the universe. However, nongravitational means are crucial in the formation of stars, galactic winds, and to the reshuffling of mass within galaxies. Also, some Copenhagen interpretation physicists believe that QM governs everything in the universe. We maintain that all laws of physics break down at the '1', that there are *no* dimensions at '1'. Time, space, matter, gravity, and energy, like an electric light, are switched off. The laws of physics that we regard as immutable, as sacred, are no more. And as there is no separation of the observer from the observed, '1' makes it possible to generate "the knowledge of ourselves" (self-identification—becoming aware of ourselves) and to increase precision based on known boundaries in '1': the fundamental physical constants and the laws of physics. [3]

Mathematically elegant phenomenological description of the electromagnetic field [24] is the unifying theme of the zero velocity frame of reference [25] as realized by J. C. Maxwell's 1873 "medium or substance in which the energy exists after it leaves one body and before it reaches the other" in electricity and magnetism. The '1' can be observed in H. A. Lorentz's transformations, A. Einstein's geometrodynamics, and J. A. Wheeler's "the boundary of a boundary is zero." Also, the empty space is rapidly emerging as the pivotal edifice of concepts of space, time, matter, quantum, hierarchy of theories of things, and relativity in the writings of I. J. R. Aitchison, M. Atiyah, P. J. Braam, D. Finkelstein, G. N. Fleming, B. J. Hiley, R. Penrose, D. W. Sciama, and other similar works. [26]

The physical relationships and the fundamental physical constants, as mentioned earlier, are intervals that characterize the foundation of physical science: the scale invariant zero velocity frame of reference, '1'. Specifically, the theories of Newton, Lorentz, Einstein, and Planck, among others, are not only about gravity, transformations, geometrodynamics, photons, and the current standard model of particle physics, but also are more significantly about the "mathematization of the language" for the description of the software structure, organization, boundary, and hierarchy in the coordinate system "at rest," '**1**'. '**1**' constitutes the ultimate nature of all phenomena. In '**1**', all principles in nature and parameters for atomic structure, algebraic topology, symmetries, and ground of being *preexist* as algorithmically compressed potentiality. [16-19]

This means that everything that is, was, and is going to happen tunnels into visible existence out of the "algorithmically compressed" starting state, '**1**'. Expressly, what we validate is a derivative, a "table lookup," of the "preexisting hidden variables" (algorithmically compressed information) of the zero velocity frame of reference. If what we validate does not preexist, in all possible states simultaneously, we can never verify it with measurements. Just as accessing information in a computer, *observation* of the observer accesses information (manifests existence) in '**1**'. To uncover which state the preexistence is in, we have to make the observation, which collapses the wave function, and the preexistence goes into a definite state. What is significant, for Bohr reality existed only after an observation was made, while for Einstein this "objective reality," or '**1**', can exist in definite states outside human experience. Thus *at* '**1**', Einstein's insight is true: "The Moon is still there [preexists] when nobody is looking at it." The '**1**' is that which, when you stop observing, doesn't go away. However for Bohr, Heisenberg, and Schrödinger's QM, to observe the moon, the wave function has to undergo a sudden change at the stationary state—a "collapse" to a new state, representing an observable quantity by the observer. When you shine a light at a stationary object, you see what is reflected.

Consequently, Heisenberg is correct for QM; it is impossible to know precisely both (simultaneous measurement of *two* distinct constants) the particle's position (where it is) and momentum (how fast it is going) with only *one* wave function "collapse" at the stationary ground state. We need *two* measurements. If you have to measure five distinct constants you need five separate measurements. If you have to measure one thousand distinct constants you need one thousand separate measurements. There is no uncertainty, as Heisenberg suggested. However, as we will see in papers 2 and 3 and as E. Wigner put it, through the unreasonable effectiveness of mathematics (and measurement) in the natural sciences, one can know precisely the *preexisting landscape* of the '**1**' that measurements cannot achieve without mathematics. Here again, *at* '**1**', Einstein is right: "God does not 'play dice' with the universe."

In mathematics the concept of the zero has been developed. The significance of the zero and its positional place value notation in numbers is universally recognized and utilized in mathematics. In the formalism of physics, the "rest frame," equivalently, the nature of the zero velocity frame of reference and its "transformational" potential, is neither universally recognized as a fixed and unchanging mathematical *and* experimental physics structure nor utilized for the prediction of qualitative physics. Namely, we do not characterize the nature of the boundary between the quantum and classical realms, the "positional place value notation of physical quantities," the hierarchy of symmetries, matrix mathematics, and the scale of measurement for the laws of physics and the fundamental physical constants.

Similar to the *empty space that integrates letters on this page*, the '1' "instantaneously" integrates all diverse phenomena in nature. Here, with "zero time delay," the phase shifts in the wave function of the electron occur, and all the laws of physics are *instantaneously* transformed. Here we find information on *how* the unifying process of group representation of phenomena should take place, together with all the space-time geometry, the guiding principle of the standard model, the evolution of the universe, and quantum information of the dynamical law. We use this "boundary of a boundary," between the physical quantities, for the unification of the scientific process as well as for the experimentally observable mathematization of physics through the LSPR discussed in paper 4. We characterize that boundary by way of the fundamental constants discussed in papers 2 and 3.

In reference [14], see Appendix I, the concept of the '1' as it relates to the positional place value notation and the experimentally observable scale of measurement for the laws of physics and the fundamental physical constants has been formulated in precise language and validated experimentally. Consequently, it will not be reconstructed at this time. For purposes of prediction, here we will comment only briefly on the zero. Each number, on a mathematical scale of numbers, has an a priori state, the *mathematical zero*, as its initial condition, or a common starting point, of the frame of reference for the "ordering of numbers":

$$-3, -2, -1, 0, 1, 2, 3. \tag{1.1}$$

We suggest that each physical quantity on the scale of quantities also has an initial condition, the zero velocity state, which in the laws of physics (i.e., $1 = E/m_e c^2$) is the *natural zero* '1'$\rightarrow$0 ($1 = 10^0 = e^0 = x^0$). The '1' is the starting point and the nature of the boundary for the foundation of the "order of the physical quantities" and the "order of the laws of physics." For relationships, see table 1.1 and the referenced constants in papers 2, natural unit of conductance, G_r, equations (2.9b), and natural unit of resistance, Z_o, equation (2.11b), and 3, natural unit of period of harmonic motion, T, equation (3.32) and natural unit of frequency, f, equation (3.31):

$$T, G_r, \text{`}\mathbf{1}\text{'}, Z_o, f. \tag{1.2}$$

In papers 2-4, we will see that '**1**' emerges as the foundation of physics. The '**1**' is so severely constrained by the requirements of mathematical and measurement consistency, with its various laws fixed by logical necessity, that it serves as the essential nature of things, the theory of everything, or the basic ensemble theory of modern physics.

3 Testable Predictions via the '1' and the One-and-the-Many Principle

The '**1**' and the one-and-the-many principle [3, 27] enable us to explain in very precise formulation the nature of the boundary conditions, '**1**', between the quantum and classical realms—between the microscopic and macroscopic worlds. This principle is analogous to Cantor's sets where he attempted to solve the different order of actual infinities (transfinite numbers) [28] and which is discussed in the "Orders of Infinity," paper 4. The foregoing is based on the fact that the *one* (an individual quantity, q_k) at equilibrium state, '**1**', is a "reciprocal" of the *many* (a number of q_k's: group quantity, as a symmetry group, and the mathematical expression of the notion of symmetry, Q_k, equivalent to H. Weyl group representation theory) at equilibrium state, '**1**', where

$$\mathbf{1} = Q_k\, q_k, \tag{1.3}$$

and q_k is either equal to or less than ($\leq$) '**1**', or '**1**' is equal to or less than ($\leq$) Qk, where

$$q_k\, (\leq)\, \mathbf{1}\, (\leq)\, Q_k. \tag{1.4}$$

The correspondence of q_k and Q_k values are determined by the fundamental physical constants that can be found in papers 2-3. The q_k or Q_k relationships can encompass a single quantity, q_k, or can be a group of

q_k quantities, Q_k. Consider this dual picture, table 1.1, of individual q_k and group Q_k phenomena in the one-and-the-many principle. Predicted new q_k as well as Q_k relationships are highlighted with bold letters:

TABLE 1.1. The '1' in the One-and-the-Many Principle

q_k: **Individual Quantity**	Q_k: **Group Quantity**
Compton wavelength λ_c	Number of waves n, where $\lambda_c n = 1$
Period of harmonic motion T	Frequency f, where $Tf = 1$
Conductance G_r	Resistance Z_o, where $G_r Z_o = 1$
Conductance quantum G_o	Inver. cond. quantum G_o^{-1}, where $G_o G_o^{-1} = 1$
Fine-structure constant α	Inverse fine-structure const. α^{-1}, where $\alpha\alpha^{-1}=1$
Magnetic flux quantum Φ_o	Josephson constant K_J, where $\Phi_o K_J = 1$
Quantized Hall conductance e^2/h	von Klitzing constant R_k, where $e^2 R_k/h = 1$
Linear time τ	Velocity (linear) v, where τv $= 1$
Penetrability of free space z_o	Univ. grav. const. G_e, where $z_o G_e = 1$
Observer	**Observed,** where observer x observed = '1'
The point, '1'	**The empty space,** '1', where '1' x '1' = '1'

In 1923, Niels Bohr fathered the correspondence principle. [29] It gave us a new theoretical basis for the development of a comprehensive equivalence linking the quantum theory and classical physics. In Bohr's coat of arms is *Cotraria sunt complementa*; it means "opposites are complementary."

In string theory, the notion of duality was observed in the radius symmetry $Rs \rightarrow 1/R_s$ of a circle known as T-duality. [30] In Feynman's "sum-over-paths" approach to QM, we see duality with the '1', where each individual electron actually traverses simultaneously (it exists there as a hidden dimension) every possible path connecting its starting position with its final objective. In consequence, only one of the infinity of paths emerges as manifest reality. 1/Rs of a circle known as T-duality. [30] In Feynman's "sum-over-paths" approach to QM, we see duality with the '1', where each individual electron actually traverses simultaneously (it exists there as a hidden dimension) every possible path connecting its starting position with its final objective. In consequence, only one of the infinity of paths emerges as manifest reality.

What have been *overlooked* in the correspondence principle, the radius symmetry, and the equations of table 1.1 are the '1' and our most elegant and rigorous avenue in the search for the one-to-one correspondence in the fundamental laws of nature, unity, continuity, nonlocality, and novel predictions. At this juncture, '1' serves as the nature of the boundary between the quantum and classical observations. Here '**1**' imposes *consistency, order,* and *renormalization* on nature. It facilitates transformation of quantities, constants, symmetries, and geometry from one level or hierarchy of physical reality to the next that produces contradictory pictures about measurements.

Notice that the physical equipment required for determining the q_k and Q_k quantities are incompatible with each other; they are mutually exclusive. The uncertainty principle is the general formulation of this duality, allowing only one out of a pair of physical values to be measured with subjective accuracy. In this sense the elucidation of the q_k and Q_k phenomena are complementary.

We have utilized this principle to predict the hierarchy of symmetries in the electric, magnetic, and gravitation relationships (table 4.1) and the physical quantities and constants (papers 2-3). Also, we have used it in the formulation of the theories of physics as seen in the magnetic constant equation (1.9a), electric constant equation (1.10a), gravitational constant equation (3.38a), the nature of time and space (below), and the natural spectrum of physical quantities and the LSPR (paper 4).

4   The Nature of Time

The development of the space-time model has proven to be a challenge in physics. A fundamental postulate of special relativity is that space and time are linked together. But so is every other physical quantity in '1'.

The equations of special relativity make the speed of light equal to one. When the speed of light is equal to one, $E = m$, energy and mass become equal. As a result, particle physicists use the same units to measure energy and mass. Wolfgang Pauli had a term for such concepts. He would express them as "it is not even false" (*Das is nicht einmal falsch*), meaning that they were so completely wrong that they could not even be utilized to make predictions to compare with observations, right or wrong.

In paper 3 we will see that in QED, c^2 of special relativity, $E = m_e c^2$, equation (3.61), is the gravitational potential V_g, equation (3.36a). We can appreciate why Schrödinger's wave mechanics version of the QM equation, for time evolution of the wave function, describing the state of the system, was experimentally inconsistent with the principles of special relativity.

According to Bohr, "a space-time description is impossible." [1] In our picture, because in 3 + 1 expression we don't know what type of time or space the topological 4-manifold represents, 3 + 1 description is not computationally classifiable. Therefore, in our description the 3 + 1 space-time is divided into space and time correspondingly. Notice, the curvature of space-time and its description by an object known as a tensor (i.e., the Riemann curvature tensor, the Weyl curvature, the Ricci curvature) can be simplified.

Utilizing '1' and the one-and-the-many principle, we suggest that *time is the smallest QED increment* (reciprocal relation) *of motion* in '1'. [27, 31] Here we see the unification of motion and rest. Incidentally, motion and rest was described by Galileo Galilei and characterized by Isaac Newton's as the first law of motion, the principle of inertia. When at rest, motion is equal to '1'.

As Q_k quantity, we differentiate three forms of motion: linear, angular, and cyclic. Thus, as q_k quantity of motion, we distinguish three quantities of time: linear, angular, and cyclic.

Linear time, τ, equation (3.53a), (= 1/linear velocity, v, equation (3.28)), is characterized in units of *seconds* per meter in '1'. Linear time is a measure of time delay in electric wires of computers *per meter* of wire or the astronomical distance that light travels in a year. The natural unit of linear time in '1' is $\tau = 3.3356 \times 10^{-9}$ (1/2.9979 x 10^8) s m^{-1}, equation (3.53b); a light-year is equivalent to 9.460 5284 x 10^{15} m s^{-1}.

Angular time, t, equation (3.3a), (= 1/angular velocity, ω, equation (3.30a)), is characterized in units of *seconds* per radian in '1'. Angular time is a rotational or curved measure of time used in our clocks. The natural unit of angular time in '1' is $t = 1.1812 \times 10^{-22}$ s r^{-1}, equation (3.3b). Present CODATA measurements of (angular) time ($t_c = \hbar/m_e c^2 = 1.288\ 088\ 6570 \times 10^{-21}$) do not include our planet, the solar system, the Milky Way galaxy, and the Local Group movements in reference to '1'. Note, when measuring the velocity of light, we reference the empty space, '1', but not with time. Paper 3, equation (3.30a), shows the derivations.

Periodic time, T, equation (3.32), (period of harmonic motion = 1/frequency, f, equation (3.31)), is described in *seconds* per cycle, Hz^{-1}. The natural unit of periodic time in '1' is $T = 8.0933 \times 10^{-21}$ s c^{-1}, equation (3.32).

Since the past, present, and future "preexist" simultaneously in '1' as the algorithmically compressed "arrow of time" potentiality, this suggests that *no motion in '1' produces no arrow of time in '1'*. From the viewpoint of quantum physics, the past is tangible in the present form of physically expressed facts. The present is the point in time when potentiality, '1', transforms into tangible physical reality. The future, '1', is fundamentally open and unequivocally undetermined. However, a field of possibilities (alternative timelines) or prediction is brought to light for whose realization positive probabilities, time's arrow, can be given. We use this field of possibilities in the prediction of physics.

Since '**1**' contains infinite possible arrangements of matter and events (i.e., parallel worlds, algorithmically compressed alternate universes), then time is simply a way to put those things (i.e., observer dependent—personal universe) into an algorithmically compressed arrow-of-time sequence in '**1**'. Thus, time advances like a movie, one frame at a time with different increments of motion in '**1**'. Time increments in '**1**' can be expressed as follows:

$$T/t = 2\pi_{qed}, \text{ equation (3.52b).} \tag{1.5}$$

$$T/\tau = \lambda_c, \text{ equation (2.8b).} \tag{1.6}$$

$$t/\tau = S, \text{ equation (3.4b).} \tag{1.7}$$

$$1/(t\tau) = g, \text{ equation (3.35b).} \tag{1.8}$$

5 The Nature of Space

Routinely we consider that space has three dimensions, 3-D. However, one can divide the 3-D space as Jules Henri Poincaré [32] and others have done. To divide the 3-D space into 2-D surfaces, 1-D lines and point cuts is essential. We cannot go further, since a point is irreducible. As with time, through '**1**', motion, and the one-and-the-many principle, one can shed additional light on the nature of space. We suggest six descriptions of space. Two are dimensionless, and four are dimensional, where the metric relations of space may be determined.

The *dimensionless* descriptions of space are the *point* and its inverse, the *dimensionless empty space*, equation (4.10). Both the point (as "one") and the dimensionless empty space (as "many") between the stars, or the particles in an atom, are nonlocal, immovable, infinite (continuing beyond the confines of our observable universe), and are characterized in the laws of physics with the symbol '**1**'. In the setting of the a priori awareness, '**1**' has neither limits nor dimensions nor the traditional geometry of Euclid, Georg Riemann, et al., involving measurement. It is a topological invariant—that is, invariant of the dimensionless empty space. Here there are no metric properties of the 3-D manifold nor dynamics (i.e., every arrangement of the dimensionless empty space satisfies the Euler-Lagrange equation). E. Witten's field theory is one about pure space. [20, 33]

The four-dimensional, or metric, relations of space are the *periodic space* and the *angular or curved space* and their reciprocal quantities. Periodic space and angular space are compound motion quantities in '**1**'. That is, pairs of motions, when integrated, tend to neutralize each other, thus producing a compound "secondary motion" that manifests as periodic or angular space in '**1**'. We can draw a parallel to complementary colors where one of a pair of colors opposed to the other element of the pair on a schematic chart or scale, as green is opposed to red, when integrated tend to neutralize each other. In a complementary angle, either of two angles that are added produces an angle of 90°.

Periodic space, as "one," q_k, quantity, is characterized in the laws of physics as wavelength λ_c in '**1**', equations (2.8a) and (2.8b), expressed in meters per cycle. Conversely, periodic space as "many," Q_k, quantity, is characterized in number of wavelengths in '**1**' ($n = 1/\lambda_c$, see table 1.1) and is expressed in cycles per meter. Periodic space consists of linear velocity, v, equation (3.28), and periodic time, T, equation (3.32), in '**1**', where $\lambda_c = vT$, equations (3.32) and (2.8a).

Angular or curved space, as "one," q_k, quantity, is characterized in physical relationships as a line, distance, or the gyroradius (sometimes called the Larmour radius) of a charged particle, S, equation (3.4a), expressed in meters per radian in '**1**'. Angular space as "many," Q_k, quantity, is characterized as a natural unit of circular wave number Λ in '**1**', equations (3.54a) and (3.54b), in radians per meter. Like periodic space, angular space is a compound motion quantity, consisting of linear velocity, v, equation (3.28), and angular time, t, equation (3.3a), in '**1**', where $S = vt$, equation (3.4a).

Also, instead of cutting the 3-D space, as Poincaré did, we can unfold it from the point, '**1**', to the 1-D space of natural unit of length, equation (3.4a). This is achieved as an interval, S, between two points in empty space, expressed as '**1**' and equality '=' in the laws of physics. [3, 27] Note that the "interval," like a number in mathematics (i.e., "meaning") between the various points in '**1**' makes measurements *and* physics possible. We utilize this interval along with the "boundary of a boundary" in '**1**' (i.e., the "initial conditions" of measurement) for the mathematization of physics and the construction of the LSPR. While the equality '=' interval, like a step, is deterministic, we suggest that at the boundary of a boundary, between the steps, the participator can employ free will. Then, we can go to the 2-D space as the natural unit of area, equation (3.5a), and the 3-D space of the natural unit of volume, equation (3.6a).

In addition, as with time so with dimensional space, the alternative space lines (alternative universes, multiverse) transform from potentiality into physical reality in the continuing unfolding of '**1**'. Specifically, the unique world of our experience emerges from the multiplicity of alternatives available in the empty space. The same applies to the gravitational field strength (i.e., g, equations (1.8) and (3.35b), where $g = 1/(t\tau)$), energy, and matter. As with time so with space, no motion in '**1**' generates no dimensional space of past, present, or future in '**1**'. And as every physical relationship requires time and space (i.e., $2\pi_{qed} = \lambda_c/S = T/t$, equations (1.11a) and (1.5)), it means that no motion in '**1**' generates any physical reality in '**1**'.

Thus, a particle, or a set of multiple possible universes, has a potentiality in '**1**' independent of the measurements. Accordingly, the 2π rotation group in a given number of spin dynamics dimensions (i.e., the atom, our planet, the solar system, the Milky Way galaxy, and the Local Group) in '**1**', which we characterize with the symbol π_{qed}, brings into being algorithmically compressed information of '**1**' that we experience and measure.

The present evidence is that the universe is expanding. That means that the mean distance between conserved particles is increasing with time in relationship to Hubble's constant. [34] This suggests that the intervals (fundamental physical constants) are constantly changing in '**1**' and were therefore different in different periods of the universe. We have noticed that the spectra of constants of thirty years ago are slightly different from the spectra of constants measured now.

Furthermore, we can consider the universal expansion in terms of the age of the universe. Namely, as time advances, the age of the universe advances at an accelerating rate *approaching potential* infinity. Potential infinity enabled mathematicians to develop the concept of a limit. Therefore, our observation and understanding of the "birthday" of the universe, the beginning of space-time expansion, will go faster backward in time to potential infinity.

6   Rigorous Prediction of Physics via the Five Constants

It is usually assumed that experiment is the only source of truth, and as Poincaré stated, "It alone can teach us something new; it alone can give us certainty." We agree with Roger Penrose et al. "that the structure of the physical world is dependent, very precisely, upon mathematics." [35] As '**1**' is algorithmically compressed mathematics, one requires but a few mechanical rules with a few verified high-precision metric relations to calculate a particular case of "extended magnitude" from '**1**' or to determine the fundamental laws governing physical phenomena.

To predict exactly the laws of physics and their precise values of the fundamental physical constants that don't have to be verified by elaborate and costly experiments in classical and QM we require, as stated earlier, scale invariance, absolute degree of consistency, order, and fundamental simplicity for the basic mathematical steps, which we will describe in more detail in papers 3 and 4. At this stage we will demonstrate how the fundamental physical constants and equations of physics can be calculated from first principles. To achieve this we utilize four independent very high-precision constants plus our calculated quantization constant.

In layman's terms the four high-precision constants, which yield a wealth of essential insight into real-world system of the Universe, can be equated to a bowed four stringed musical instrument of the violin family (violin, viola, cello, and double bass) while the "renormalizable" merry-go-round *quantization constant* that "has a length but no other dimension" can be related to the violin family *bow*.

In the standard symphony orchestra, traditional string quartet, and other ensembles, the string player pulls the bow across the strings of a musical instrument causing string oscillation and movement of the air around it, producing sound waves. In technical terminology of physics, as time passes, each oscillation mode of the four high-precision constants gives rise to principles of symmetry that dictate the dramatis personae of the drama, principles of condensed matter physics, different species of quantities and particles with its mass, charge, volume, density, gravity, plus an infinite variety of properties determined by the quantization constant.

To enter the labyrinth of explanation through a different door that describes the fundamental simplicity of the four independent high-precision constants are the four chambers of the heart (right atrium, right ventricle, left atrium, and left ventricle) while the quantization constant can be viewed as the merry-go-round frame of reference with different pumping states of the heart. Equally we can shed light on the four constants by way of the four seasons (Spring, Summer, Fall, and Winter) while the suggested quantization constant can be considered by way of the merry-go-round cyclic states of the laws of nature.

Please be so kind and forgive our humble reach for the '**1**' and criticism toward ourselves in the esoteric theory of strings. Our work reveals that some string and M-theorists venture to produce observational evidence of "music" (i. e., calculations and *measurable predictions* for the foundation of universal physical law that are meant to describe the nature of ultimate and relative reality) without utilizing the fundamental physical constant *reference*. As a result their one-dimensional quantization string turns out to be inadequate in the prediction of different species of condensed matter physics, quantities, or the fundamental physical constants. It's like a violin, cello, or modern double bass bowing for music that has no strings, or writing with a hand which has only the thumb, or a pulsating human heart which is short of its four chambers of the heart.

The four high-precision constants are the following: (1) The permeability of free space (magnetic constant), where standard uncertainty is exact and relative standard uncertainty is exact. (2) The permittivity of free space (electric constant), where standard uncertainty is exact and relative standard uncertainty is exact. (3) The Rydberg constant, where standard uncertainty is 0.000 055 m^{-1} and relative standard uncertainty is 5.0×10^{-12}, and (4) the elementary charge, where standard uncertainty is $0.000\ 000\ 035 \times 10^{-19}$ C and relative standard uncertainty is 2.2×10^{-8}. The four constants are from the Committee on Data for Science and Technology (CODATA) of the International Council for Science, with internationally suggested values of the basic constants and conversion factors of physics. The "2006 and 2010 CODATA recommended values" were achieved under the auspices of the CODATA Task Group on Fundamental Constants that are available online in a searchable database provided by the NIST fundamental constants data center. The URL is http://physics.nist.gov/constants. This new set of constants became available in March 2007 and replaced the 2002 [36] CODATA values. A detailed description of the constants is in preparation and will be posted on arXiv.org and submitted for publication.

Contrasting the necessity of high-precision measurements in fundamental physical constants, the "mathematical constant 2π" is *defined independently of any physical measurement* in QED. It is usually assumed, [36, 37] without rigor of treatment, precision testing, and measurement protocol (which can specify a basis of the rotation group spin space) that in QED, constant 2π has the same geometry, description of spin dynamic evolution, and value not only in classical physics, but also in atomic and nuclear physics. [38, 39] However, we discovered that since the electron and other elementary systems (i.e., proton, neutron) in motion have different geometry [40] and value than the circle, the mathematical constant 2π in QED can have physical subtleties [41], geometry, and value that are very different (i.e., it is 10.9049 times larger) from the accepted transcendental 2π number. Please note, for problems in physics, once a high source of error is introduced to fundamental constants (i.e., quantum anomaly, anomalously highly interactive states of matter, the Heisenberg uncertainty principle) great confusion, uncertainty, and excitement suddenly develop in high-precision measurements and dimensional uniformity of physical equations. Examples: conformal anomaly (anomaly of scale invariance), gravitational anomaly (also known as diffeomorphism anomaly), parity anomaly, chiral anomaly, global anomaly, etc.

The velocity of light is one of the most important of the fundamental constants of nature. For dimensional uniformity of physical equations, when in QED, we generally reference the velocity of light to the coordinate system at rest. Similarly, when in QED, we *must* reference the spin dynamic rotation group of the mathematical constant 2π to the coordinate system at rest. Please observe that the path of the orbits of the electron, proton, neutron, our planet, the solar system, the Milky Way galaxy, and the Local Group of the rotation group in the given number of dimensions *is* in reference to the coordinate system "at rest," '1'. Our suggested quantization constant, $2\pi_{qed}$, is a derived "electron 2π" value [2] from the fine-structure constant, $\alpha = 7.297\,352\,5376$ (50) x 10^{-3}, that was also obtained from CODATA. For particulars on the suggested quantization formalism, please see paper 3.

To spell out for the reader what is old and what is new in the equations listed in papers 1-4, we have identified the following:

a. Predicted *new physical relationships* with the symbol {$}
b. Predicted *new fundamental physical constants* with the symbol {#}
c. Predicted *old fundamental physical constants with "higher precision"* than the 2006 CODATA recommended values with the symbol {*}
d. 2006 CODATA recommended values with the symbol {**}
e. 2010 CODATA recommended values with the symbol {***}
f. Predicted *old fundamental physical constants with "higher precision"* than the 2010 CODATA recommended values with the symbol {****}

In equations (1.9a)-(1.13a), we give relationships for equations (1.9b)-(1.13b) constants. Physical constants of these relationships are cross-referenced to papers 2-3. The four CODATA constants used are exhibited in equations (1.9b), (1.10b), (1.12b), and (1.13b). Our suggested new quantization constant is shown in equation (1.11b).

The permeability of free space (magnetic constant):

$$\mu_o = L/S = LR_\infty(4\pi_{qed})^3; \{\$\}. \tag{1.9a}$$

$$\mu_o = 4\pi \times 10^{-7} = 12.566\ 370\ 614\ 359\ 173 \cdots \times 10^{-7}\ \text{N A}^{-2}.\ \{**\} \tag{1.9b}$$

$$\mu_o = 4\pi \times 10^{-7} = 12.566\ 370\ 614\ 359\ 173 \cdots \times 10^{-7}\ \text{N A}^{-2}.\ \{***\} \tag{1.9b}$$

Relative standard uncertainty: exact. Standard uncertainty: exact. For L relations, see equation (3.8a); S, see equation (3.4a); R_∞, see equations (1.12b) and (3.57); and π_{qed}, see equations (1.11a) and (3.52a). [2, 31]

The permittivity of free space (electric constant):

$$\varepsilon_o = C/S = CR_\infty(4\pi_{qed})^3; \{\$\}. \tag{1.10a}$$

$$\varepsilon_o = 1/\mu_o c^2 = 8.854\ 187\ 817\ 620\ 389 \cdots \times 10^{-12}\ \text{F m}^{-1}\{**\} \tag{1.10b}$$

$$\varepsilon_o = 1/\mu_o c^2 = 8.854\ 187\ 817\ 620\ 389 \cdots \times 10^{-12}\ \text{F m}^{-1}\{***\} \tag{1.10b}$$

Relative standard uncertainty: exact. Standard uncertainty: exact. For C, see equation (3.7a); c = v = 299 792 458, see (****) (1.10b) and equation (3.28).

Suggested new quantization constant, π_{qed}:

$$2\pi_{qed} = \lambda_c/S = (\alpha^{-1})/2; \{\$\}. \tag{1.11a}$$

$$\{2006\}\ \pi_{qed} = 34.258\ 999\ 919\ 74\ \text{radian/cycle}; \{\#\}. \tag{1.11b}$$

$$\{2010\}\ \pi_{qed} = 34.258\ 999\ 768\ 6\ \text{radian/cycle}; \{\#\}. \tag{1.11b}$$

Relative standard uncertainty: 6.8×10^{-10}. Standard uncertainty: 0.000 000 050 x 10^{-3}. For λ_c, see equation (2.8a).

The Rydberg constant:

$$R_\infty = 1/S(4\pi_{qed})^3; \{\$\}. \tag{1.12a}$$

$$\{2006\}\ R_\infty = 10\ 973\ 731.568\ 527\ (73)\ \text{m}^{-1}.\ \{**\} \tag{1.12b}$$

$$\{2010\}\ R_\infty = 10\ 973\ 731.568\ 539\ (55)\ \text{m}^{-1}.\ \{***\} \tag{1.12b}$$

The digits in parentheses indicate the uncertainty in the last listed digits of the measurement. 2006 relative standard uncertainty: 6.6 x 10⁻¹². Standard uncertainty: 0.000 073 m⁻¹. For R_∞, see equation (3.57a); m_e, see equation (2.1a).
The elementary charge:

$$e = bS = b/R_\infty(4\pi_{qed})^3; \{\$\}. \tag{1.13a}$$

$$\{2006\}\ e = 1.602\ 176\ 487\ (40) \times 10^{-19}\ \text{C}.\ \{^{**}\} \tag{1.13b}$$

$$\{2010\}\ e = 1.602\ 176\ 565\ (35) \times 10^{-19}\ \text{C}.\ \{^{***}\} \tag{1.13b}$$

2006 relative standard uncertainty: 2.5 x 10⁻⁸. Standard uncertainty: 0.000 000 040 x 10⁻¹⁹ C. For *e* relations, see equation (3.15); for *b*, see equation (3.42).

7 Results

With '1', very severe constraints of scale invariance and the five high-precision constants, we have determined agreement between theory and CODATA observation at a level of nine to fifteen decimal places. In papers 2-3, we show the new fundamental physical constants along with CODATA values and their physical relationships. Our predicted constants (i.e., the natural unit of mass, the Planck constant, the Josephson constant, the von Klitzing constant, the Compton wavelength, atomic unit of velocity, atomic unit of energy, quantum of circulation, conductance quantum, etc.) *have been generated with the five constants*, equations (1.9b)-(1.13b). They have a higher precision that is one order of magnitude better than the "CODATA recommended values."

In addition, with the five independent basic constants, we predict fifty new fundamental physical constants that the CODATA do not provide. Also, with these five constants, we predict the "natural spectrum of physical quantities," the nature of matter, the Planck constant, the "order of the laws of physics," and five new gravitational quantities, their physical relationships, the values of their QED constants, and the LSPR.

Similar to a mathematical slide rule that is used for the prediction, verification, and a reality check of the relationships of mathematical numbers, the LSPR is utilized for discoveries of fundamental importance in scientific research, education, and the mathematization of physics. Specifically, it can be used for accelerating, empowering, and advancing the future course of theoretical and experimental physics in the prediction and validation of the fundamental laws of nature and paradigm-shifting discoveries.

Note, the physical constants (1.9b), (1.10b), (1.11b), (1.12b) and (1.13b) are based on the 2006 and 2010 self-consistent set of values of the fundamental constants and conversion factors of physics and chemistry recommended by the Committee on Data for Sciences and Technology (CODATA) for international use. However, given that *our universe is expanding*, while the laws of physics remain unchanging, the fundamental constants are no longer constant. With the expansion (as a function of red shift has shown) and the dimensionless acceleration (or deceleration) parameter of the universe the fundamental physical constants are nontrivial and are changing, as the '**1**' and the One-and-the-Many Principle (1.3) illustrates. We can use this (1.3) fundamental constants change to determine principles and equation parameters in modern physical cosmology.

For example, on March 27, 2012, the National Institute of Standards and Technology, Gaithersburg, Maryland 20899-8420, USA (Peter J. Mohr, Barry N. Taylor, and David B. Newell) recommended new 2010 values of the fundamental physical constants. The 2010 series replaces the previously recommended 2006 CODATA set and may also be found on the World Wide Web. At this juncture you will also see that because our (1.9b), (1.10b), and (1.12) are of higher precision than the CODATA recommended standards, for constants listed in (2.1a)-(2.19), in addition to demonstrating the value of our experimentally supported structure, in this way you can also validate the quality of the CODATA experimental precision. Our assessment: the 2010 CODATA measurements are exceptionally of higher precision than the 2006 high precision measurements.

Therefore, in your future derivation of new values for the basic constants and conversion factors of physics and chemistry please consult the most up-to-date CODATA values for (1.11b), (1.12b), and (1.13b) on the World Wide Web at physics.nist.gov/constants. Please notice, that the permeability of free space (magnetic constant) (1.9b) and the permittivity of free space (electric constant) (1.10b), as with the laws of physics in this reality, will remain unchanging.

References

[1] Bohr N H Archive, Copenhagen.

[2] Bedrij O (1994) Connection of π with the Fine Structure Constant *Proc Inst Math Natl Acad Sci Ukraine* N 10.

[3] Bedrij O (2002) New Relationships and Measurements for Gravity Physics. Fourth Inter Conf: Symmetry in Nonlinear Math Phys. *Proc Inst Math Natl Acad Sci Ukraine* vol. 43, Part 2, pp. 589-601.

[4] Penrose R. (1991) 'The Mass of the Classical Vacuum' in Saunders S and Brown H R, eds *The Philosophy of Vacuum* (Oxford University Press, Oxford) p 21.

[5] Neumann J von (1932) *Mathematische Grundlagen der Quantenmechanik* (Springer, Berlin); English tr by Beyer T (1955) *Mathematical Foundations of Quantum Mechanics* (Princeton University, Princeton, NJ).

[6] Bohr N (1963) *Quantum Physics and Philosophy* (Wiley, New York, NY).

[7] Wheeler J A and Zurek W H, eds (1983) *Quantum Theory and Measurement* (Princeton University Press, Princeton, NJ).

[8] Jung C G and Pauli W (1955) *The Interpretation and Nature of the Psyche*, tr by Hull R F C and Silz P (Pantheon, New York, NY).

[9] Heisenberg W (1958) *Physics and Philosophy: The Revolution in Modern Science* (Harper & Row, New York, NY).

[10] Wigner E (1967) Explaining Consciousness. *Science* 156, 798-9.

[11] Bohm D and Peat D (1987) *Science, Order, and Creativity* (Bantam Books, New York).

[12] Feynman P R (1985) *QED: The Strange Theory of Light and Matter* (Princeton University Press, Princeton, NJ) p 19.

[13] Dirac P-A-M (1949) Forms of Relativistic Dynamics. *Rev Mod Phys* 21: 392-9.

[14] Bedrij O (1993) Scale Invariance, Unifying Principle, Order and Sequence of Physical Quantities and Fundamental Constants. *Proc Inst Math Natl Acad Sci Ukraine*; also, (1994) *Ukrainian Mathematical Journal*, tr of the Proc Inst Math Natl Acad Sci Ukraine (Allerton Press, New York, NY), pp 67-73.

[15] Bedrij O (1962) Carry-Select Adder *IRE Trans Electronic Computers*, vol EC-11, pp 340-346.

[16] Bohm D (1952) A Suggested Interpretation of the Quantum Theory in Terms of 'Hidden' Variables, I. *Phys Rev* **85**, 166-179.

[17] Bohm D (1952) A Suggested Interpretation of the Quantum Theory in Terms of 'Hidden' Variables, II. *Phys Rev* **85**, 180-193.

[18] Heisenberg W (1930) *The Physical Principles of the Quantum Theory* (Dover Publications, Inc., New York, NY).

[19] Pagels H (1994) *The Cosmic Code* (Penguin Books, London).

[20] Witten E (1988) Topological Quantum Field Theory. *Comm Math Phys* 117: 353-86.

[21] Finkelstein D (1991) 'Theory of Vacuum' in Saunders S and Brown H R, eds *The Philosophy of Vacuum*. (Oxford University Press, Oxford).

[22] Everett H (1957) Relative State Formulation of Quantum Mechanics. *Reviews of Modern Physics* vol 29, pp 454-462.

[23] Garriga J and Vilenkin A (2001) Many Worlds in One. *Phys Rev D* 64, 043511, 4.

[24] Lorentz H A (1935-39) *La théorie électromagnétique de Maxwell et son application aux corps mouvants*. Reprinted in Collected Papers (Nijhoff, The Hague), vol 2, pp 63-343.

[25] Maxwell J C (1965) *The Scientific Papers of James Clerk Maxwell*, 2 vols, Niven W D, ed (Dover, NY).

[26] Saunders S and Brown H R, eds (1991) *The Philosophy of Vacuum* (Oxford University Press, Oxford) p 21.

[27] Bedrij O (2000) Revelation and Verification of Ultimate Reality and Meaning through Direct Experience and the Laws of Physics. *Ultimate Reality and Meaning, University of Toronto Press,* 23:1.

[28] Cantor G (1955) *Contributions to the Founding of the Theory of Transfinite Numbers*, ed Jourdain P (Dover, NY, 1955).

[29] Bohr N (1935) *Can Quantum-Mechanical Description of Physical Reality be Considered Complete? Phys Rev* **48**, 696-702.

[30] Glazunov N (2002) Mirror Symmetry: Algebraic Geometric and Lagrangian Fibrations Aspects, Fourth Inter. Conf.: Symmetry in Nonlinear Mathematical Physics, *Proc Inst Math Natl Acad Sci Ukraine* vol. 43, pp. 623-628.

[31] Bedrij O (1993) Fundamental Constants in Quantum Electrodynamics. *Proc Inst Math Natl Acad Sci Ukraine,* 3.

[32] Poincaré N (1904) *LaValeur de la Science* (Flammarion, Paris).

[33] Atiyah M (1991) 'Topology of the Vacuum' in S. Saunders S and H. R. Brown H R, eds *The Philosophy of Vacuum*. (Oxford University Press, Oxford).

[34] Peebles P J E (1993) *Principles of Physical Cosmology* (Princeton University Press, Princeton, NJ), p 5.

[35] Penrose R with Shimony A, Cartwright N, and Hawking S (1997) *The Large, the Small and the Human Mind* (Cambridge University Press, Cambridge, UK), p 50.

[36] Mohr P J and Taylor B N (2005) CODATA Recommended Values of the Fundamental Physical Constants: 2002. *Rev Mod Phys* vol. 77(1), pp. 1-107.

[37] Dirac P (1967) *The Principles of Quantum Mechanics* (Clarendon Press, Oxford).

[38] Dürr H, Heisenberg W, Mitter H, Yamazaki K (1959) To the theory of elementary particles *Zs Naturforch* **14A**, 441-465.

[39] Dirac P (1928) The quantum theory of the electron *Proc R Soc Lond* **117A**, 610-624; **118A**, 351-361.

[40] Dubrovin B A, Novikov S P, Fomenko A T (1990) *Modern Geometry: Methods and Applications* (Springer, New York, NY), Vols. 1-3.

[41] Fushchych, W, Zhdanow R (1997) *Symmetries and Exact Solutions of Nonlinear Dirac Equations* (Mathematical Ukraina Publisher, Kyiv).

2

Testable Prediction and Verification of the Known Fundamental Physical Constants

Believe nothing just because a belief is generally held. Believe nothing because it is said in ancient books. Believe nothing just because it is said to be of divine origin. Believe nothing because someone else believes it. Believe only what you yourself test and judge to be true.

—Siddhartha Gautama,
563-479 BC

Abstract

In paper 2, without in any way disturbing a system, we predict and verify the new physical relationships and improve the precision of the known fundamental physical constants by another order of magnitude. Specifically, based on the '1', four high-precision constants (the magnetic and electric constants, the Rydberg constants, and the elementary charge) and on our suggested quantization constant, π_{qed} = 34.258 999 919 74 radians/cycle, precise predictions of the known and new fundamental physical constants and the laws of physics are determined.

Here, for illustration purposes, we have calculated fifteen fundamental physical constants and conversion factors of physics. The derived constants include the electron mass, the Planck constant, natural unit of energy, Hartree energy, atomic unit of momentum, conductance quantum, Josephson constant, von Klitzing constant, Compton wavelength, atomic unit of electric potential, and quantum of circulation, among others. Our predicted constants are at a level of nine to fifteen decimal places and have a higher precision, by a factor of ten or more, than the "2006 CODATA recommended values."

1 Introduction

The notion of reproducible experiments, a foundation of science, is based on the concept that the universe operates according to invariant laws. The proof for the soundness of this premise is the fruition of the scientific method. Although answers have to be sought and ideas tested by experiment, at present the common experimental and mathematical formulations appear to lead to myriad challenges in numerous areas of physics. [1, 2, 3, 4] In our papers, we show that the physical constants, when scale invariant, are reducible to mathematical relations and operations that can take starkly different routes to the same physics. Based on the nonlocal hidden variables theory [5, 6] and scale invariance [7] of the '1' (i.e., $1 = E/m_e c^2$), it is possible "without disturbing the system to predict with certainty" the qualitative content of the physical constants, symmetries, equations, and continuous description of processes using high-performance computing. Essentially, the fundamental physical constants can be calculated from first principles. By using only *five* high-precision constants, equations (1.9b)-(1.13b), simple algebra, and some computational power, we have calculated QED CODATA values of constants, atomic units, and the laws of physics. In addition, with these five constants, we have calculated new constants, their relationships, and the order of the quantities, the "natural spectrum of physical quantities," and the LSPR that CODATA do not provide. [8]

Had the major players—π_{qed}, R_∞, and e, equations (1.11b)-(1.13b)—been exact, like μ_o and ε_o, equations (1.9b) and (1.10b) all predicted values of the constants would also be exact. Of course, these can easily be checked directly with the characteristic impedance of vacuum, equation (2.11a); the natural unit of conductance, equation (2.9a); the natural unit of velocity, equation (3.28); the natural unit of (linear) time, equation (3.53a); and the natural unit of gravitational potential, equation (3.36a). These are determined via μ_o and ε_o. They are exact. We have calculated them to fifteen decimal places.

As stated earlier, we are utilizing a mere five constants for the derivation and the mathematization of physics and to correct "natural" unit classifications (i.e., the natural unit of time, equation (3.3b); the natural unit of length, equation (3.4b); and the natural unit of action, equation (3.22b). [9] This methodology, arising from the '**1**', can significantly improve resolution and refine discovery of the hidden variables. It can bring more order to physics, lead to better dissemination of foundational data, and promote research excellence. Also, it can enhance, simplify, and verify the measurement and experimental processes in physics, [10] while accelerating the application of the knowledge of physics in the fields of science, technology, and education, as well as their application to human welfare. [11]

2 How Does a Unique Solution from the Multiplicity Emerge?

We begin with thirty-nine quantities. [7, 8] Applying '**1**' and the one-and-the-many principle, as described in paper 1, we determined their reciprocals. The reciprocal formula enabled us to generate twenty new physical quantities. Using a computer, we constructed a matrix with quantities in rows and columns of linear equations associated with the different routes from an initial state of '**1**'. This yielded a tremendous number of solutions and creates different kinds of environments—over one hundred thousand four—and five-term equations that unfold into a multiverse of quadrillions as the number of terms is increased.

Next, applying equations (1.9b)-(1.13b) constants, from paper 1, with different exponents, cube, and higher roots, we determined which algorithm in the '**1**', like the wave function of the system, [12] produces equivalence in the matrix relations. We noticed that each quantity can be generated many different ways. For example, mass has over one thousand relations, while the Planck constant has over 250 relations. In paper 2 we concentrated on increasing the precision of CODATA values by another order of magnitude via equations (1.9b)-(1.13b) constants. In equations (2.1)-(2.19), we predict fifteen constants with higher precision that are one order of magnitude to three orders of magnitude better than those of CODATA measurements. In paper 3 we focused on upscaling CODATA experimentally observable attainment by calculating new constants via equations (2.1)-(2.19) and the laws of physics that CODATA does not provide.

The higher-precision constants are the natural unit of mass, equation (2.1b); the Planck constant, equation (2.2b); natural unit of energy, equation (2.3b); atomic unit of energy, equation (2.4b); natural unit of momentum, equation (2.6b); atomic unit of momentum, equation (2.7b); Compton wavelength, equation (2.8b); conductance quantum, equation (2.10b); natural unit of electric potential, equation (2.13b); atomic unit of electric potential, equation (2.14b); the Josephson constant, equation (2.15b); the von Klitzing constant, equation (2.16b); quantum of circulation, equation (2.17b); magnetic flux quantum, equation (2.18b); and atomic unit of velocity, equation (2.19). In addition, using the five constants, we determined the natural unit of conductance, equation (2.9a), and characteristic impedance of vacuum, equation (2.11a). These constants have the same precision as CODATA values.

With respect to constant calculations, one should note that all are carried out with more digits than are exhibited in the text in order to avoid rounding errors. Constants of equations (2.8b), (2.10b), (2.12b), (2.15b)-(2.17b) have been calculated to eleven decimal places; underlined digits show that π_{qed} is not included in the twelfth and thirteenth decimal positions. Also, equations (2.1b)-(2.7b) and equations (2.13b)-(2.15b) have been calculated to nine decimal places; underlined digits show that the e is not included in the tenth and eleventh decimal positions. For comparison, we have included CODATA constants.

Notice every relationship of these five constants, equations (1.9a)-(1.13a), is embodied in a rigid mathematical structure of '1' based on the simple underlying principle of π_{qed}, equations (1.11a) and (3.52a), properties in physics. Since the five constants generate all other constants in QED, the mathematical structure and predictive power of π_{qed}, which we suggest embraces J. Schwarz-M. Green-E. Witten string theory and has Rosetta Stone portability to all physics. Specifically, they provide a real-world illustration that can be carried over from one physical setting to another in physics.

To realize ever more refined formalism, order, and the mathematization of '1', we apply the correspondence principle using the classical expression for all relationships, according to which the calculations of the quantum theory must correspond to the calculations of the classical theories of physics. Additionally, based on the '1', we recognize that the laws of physics are independent of the size of the physical objects being described. [13, 14] Therefore, the principle concepts, physical relationships, and constants of macrophysics (such as volume, density, etc.) are also valid in microphysics (electron, proton, etc.). [7]

3 Predictions of New Relationships and Known Constants

In the prediction of the fundamental physical constants and the laws of physics, we regard all constants and the laws of physics as determinants in the solution of simultaneous equations of the empty space. [7] Equivalent to mathematics, we apply what is known (measured and verified) and by way of the laws of physics solve for what is unknown. By knowing a few constants, at any expansion time of the universe one can predict a multitude of other constants, an endless hierarchy of relationships and quantum states for that setting of '**1**'. We have chosen equations (1.9b)-(1.13b) constants [8] because of their very high precision. A different combination is feasible. However, not every five-term constant assemblage will yield the whole spectrum of constants and symmetries. We also discovered that to describe the full spectrum of physics, five is the minimum number of constants required at present. In time, we expect the five constants will be reduced to one constant that will enable us to predict the entire spectrum of physical quantities and the fundamental physical constants.

We have included the 2006 CODATA internationally recommended values of the fundamental constants for comparison. The URL is http://physics.nist.gov/constants. Please note the CODATA Task Group on Fundamental Constants has done a very remarkable job in their integration and adjustment of multitudinous measurements from around the world. However, because our constants are of higher precision than the CODATA recommended values, in addition to demonstrating the efficacy of our experimentally supported framework, we are also corroborating the quality of the CODATA values.

4 PredictionResults

Natural unit of mass (electron mass):

$$m_e = R_\infty \mu_0 e^2 (4\pi_{qed})^3; \{\$\}. \tag{2.1a}$$

$$\{2006\}\ m_e = 9.109\ 382\ 143\ \underline{97} \times 10^{-31}\ \text{kg};\ \{*\}. \tag{2.1b}$$

$$\{2006\}\ m_e = 9.109\ 382\ 15\ (45) \times 10^{-31}\ \text{kg};\ \{**\}. \tag{2.1b}$$

$$\{2010\}\ m_e = 9.109\ 382\ 91\ (40) \times 10^{-31}\ \text{kg};\ \{***\} \tag{2.1b}$$

$$\{2010\}\ m_e = 9.109\ 382\ 910\ 25 \times 10^{-31}\ \text{kg};\ \{****\} \tag{2.1b}$$

2006 {**} CODATA value: m_e = 9.109 382 15 (45) x 10^{-31} kg. The CODATA value and our calculations agree to seven decimal places. We show one order of magnitude higher precision than the 2006 and 2010 CODATA recommended values. It should be noted that pending convergence of two experiments, according to the International Committee for Weights and Measures, the kilogram standard, along with the ampere standard, could be characterized in terms of fundamental constants as early as 2011. In equations (2.1a) and (2.1c)-(2.1f), we show how the kilogram definition can be redefined in terms of fundamental constants. Also, connecting the ampere, equation (3.1), to the elementary charge, equations (2.9a) and (2.13a), frees it from reliance on the kilogram. Now, the ampere is defined in terms of the force, in kg m/s². Additionally, "tying the kilogram to h, would be a boon for electrical measurements," says NIST's Barry N. Taylor in April 2006 *Physics Today*. In equations (2.1e), (2.1f), and (2.2a), we see this connection. Furthermore, the Josephson constant $2e/h$ and the von Klitzing constant h/e^2 can be realized in exact electrical units as shown in equations (2.15a) and (2.16a).

It may be noted that by making, equation (2.1a), $R_\infty\ (4\pi_{qed})^3 = 1/S$, as in equation (3.4a), electron mass m_e can be regarded as a "pair" of electric charges e, equation (1.13b), in permeability of free space μ_o, equation (1.9b), per unit of length S:

$$m_e = \mu_o e^2/S;\ \{\$\}. \tag{2.1c}$$

Similarly, we can consider m_e as a "pair" of magnetic fluxes ϕ, equation (3.16), in the permittivity of free space ε_o, equation (1.10b), per unit of length S:

$$m_e = \varepsilon_o \phi^2/S;\ \{\$\}. \tag{2.1d}$$

In addition, we can attain the marriage of light and matter in QED via the electron mass, m_e, equations (2.1e) and (2.1f), with the Planck constant, as pairs of electric charges or magnetic fluxes in h, equation (2.2a):

$$m_e = \mu_o e^2 (2\pi_{qed})/\lambda_c;\ \{\$\}. \tag{2.1e}$$

$$m_e = \varepsilon_o \phi^2 (2\pi_{qed})/\lambda_c;\ \{\$\}. \tag{2.1f}$$

Planck constant:

$$h = \mu_o e^2 (2\pi_{qed}) v = \varepsilon_o \phi^2 (2\pi_{qed}) v; \{\$\}. \tag{2.2a}$$

$$\{2006\}\ h = 6.626\ 068\ 959\ \underline{27} \times 10^{-34}\ \text{J s}; \{*\}. \tag{2.2b}$$

$$\{2006\}\ h = 6.626\ 068\ 96 \times 10^{-34}\ \text{J s}; \{**\}. \tag{2.2b}$$

$$\{2010\}\ h = 6.626\ 069\ 57\ (29) \times 10^{-34}\ \text{J s}; \{***\}. \tag{2.2b}$$

$$\{2010\}\ h = 6.626\ 069\ 575\ 2 \times 10^{-34}\ \text{J s}; \{****\} \tag{2.2b}$$

2006 {**} CODATA value: $h = 6.626\ 068\ 96 \times 10^{-34}$ (33) J s. Where the natural unit of velocity in QED v = *c*. Here, the CODATA value and our calculations agree to seven decimal places. We show higher precision that are one order of magnitude better than those of the 2006 and 2010 CODATA recommended constants.

We know that when a photon hits a material, it can emit only one and only one electron. Similarly, in the transformation of m_e into h, please notice $\mu_o e^2$ of equations (2.1c)/(2.1e) and $\varepsilon_o \phi^2$ of equations (2.1d)/(2.1f) in equation (2.2a). In equation (2.2a), $\mu_o e^2$ and $\varepsilon_o \phi^2$ break free (quantum-mechanically tunnel away) from an atom to become h. They are no longer "chained" in the $1/S$ or $(2\pi_{qed})/\lambda_c$ atom's Coulomb well as m_e, equations (2.1c)/(2.1e) and (2.1d)/(2.1f), but have overcome the electrostatic attraction to climb out of the atom in equation (2.2a). One can achieve this result with powerful electric fields from a laser pulse, as was recently demonstrated by a German-Dutch team headed by Ferenc Krausz of the Max Planck Institute for Quantum Optics in Garching, Germany, and described in the April 5, 2007, *Nature*.

Natural unit of energy (electron mass energy equivalent):

$$E = R_\infty e^2 (4\pi_{qed})^3 / \varepsilon_o; \{\$\}. \tag{2.3a}$$

$$\{2006\}\ E = 8.187\ 104\ 376\ \underline{99} \times 10^{-14}\ \text{J}; \{*\}. \tag{2.3b}$$

$$\{2006\}\ E = 8.187\ 104\ 38 \times 10^{-14}\ \text{J}; \{**\}. \tag{2.3b}$$

$$\{2010\}\ E = 8.187\ 105\ 06\ (36) \times 10^{-14}\ \text{J}; \{***\}. \tag{2.3b}$$

$$\{2010\}\ E = 8.187\ 105\ 065\ 69 \times 10^{-14}\ \text{J}; \{****\}. \tag{2.3b}$$

2006 {**} CODATA value: $m_e c^2$ = 8.187 104 38 x 10^{-14} (41) J. Here, the CODATA onstant and our calculations agree to seven decimal places. We show higher precision by a factor of ten than the 2006 and 2010 CODATA recommended values.

Atomic unit of energy:

$$E_h = E/(4\pi_{qed})^2; \ \{\$\}. \tag{2.4a}$$

$$\{2006\} \ E_h = 4.359\ 743\ 938\ \underline{89} \times 10^{-18}\ J; \ \{*\}. \tag{2.4b}$$

$$\{2006\} \ E_h = 4.359\ 743\ 94 \times 10^{-18}\ J; \ \{**\}. \tag{2.4b}$$

$$\{2010\} \ E_h = 4.359\ 744\ 34\ (19) \times 10^{-18}\ J; \ \{***\}. \tag{2.4b}$$

$$\{2010\} \ E_h = 4.359\ 744\ 344 \times 10^{-18}\ J; \ \{****\}. \tag{2.4b}$$

2006 {**} CODATA atomic unit of energy = 4.359 743 94 x 10^{-18} (22) J. In atomic unit of energy, the CODATA value and our calculations agree to seven decimal places. We show higher precision by a factor of ten than the 2006 and 2010 CODATA endorsed constants. Note, when the atomic unit of energy, E_h, equation (2.4a), is multiplied by $(4\pi_{qed})^2$ quantization constant, the product is the natural unit of energy E, equation (2.3a).

Hartree energy:

$$E_h = E/(4\pi_{qed})^2; \ \{\$\} \tag{2.5a}$$

$$\{2006\} \ E_h = 4.359\ 743\ 938\ \underline{89} \times 10^{-18}\ J; \ \{*\}. \tag{2.5b}$$

$$\{2006\} \ E_h = 4.359\ 743\ 94 \times 10^{-18}\ J; \ \{**\}. \tag{2.5b}$$

$$\{2010\} \ E_h = 4.359\ 744\ 34\ (19) \times 10^{-18}\ J; \ \{***\}. \tag{2.5b}$$

$$\{2010\} \ E_h = 4.359\ 744\ 344\ 1 \times 10^{-18}\ J; \ \{****\}. \tag{2.5b}$$

2006 {**} CODATA value: E_h = 2 $R_\infty hc$ = 4.359 743 94 x 10^{-18} (22) J. Here, the CODATA constant and our calculations agree to seven decimal places. We exhibit higher precision that is an order of magnitude better than that of the 2006 and 2010 CODATA values.

Natural unit of momentum:

$$p = R_\infty v \mu_o e^2 (4\pi_{qed})^3; \{\$\}. \qquad (2.6a)$$

$$(2006)\ p = 2.730\ 924\ 063\ 80 \times 10^{-22}\ kg\ m\ s^{-1}; \{*\}. \qquad (2.6b)$$

$$(2006)\ p = 2.730\ 924\ 06 \times 10^{-22}\ kg\ m\ s^{-1}; \{**\}. \qquad (2.6b)$$

$$(2010)\ p = 2.730\ 924\ 29\ (12) \times 10^{-22}\ kg\ m\ s^{-1}; \{***\}. \qquad (2.6b)$$

$$(2010)\ p = 2.730\ 924\ 293\ 53 \times 10^{-22}\ kg\ m\ s^{-1}; \{****\}. \qquad (2.6b)$$

2006 (**) CODATA value: $m_e c = 2.730\ 924\ 06 \times 10^{-22}$ kg m s^{-1}. Here, the CODATA value and our calculations agree to eight decimal places. We provide higher precision by a factor of ten than the 2006 and 2010 CODATA recommended values.

Atomic unit of momentum:

$$\hbar/a_o = p/(4\pi_{qed}); \{\$\}. \qquad (2.7a)$$

$$(2006)\ \hbar/a_o = 1.992\ 851\ 564\ \underline{69} \times 10^{-24}\ kg\ m\ s^{-1}; \{*\}. \qquad (2.7b)$$

$$(2006)\ \hbar/a_o = 1.992\ 851\ 565 \times 10^{-24}\ kg\ m\ s^{-1}; \{**\}. \qquad (2.7b)$$

$$(2010)\ \hbar/a_o = 1.992\ 851\ 740 \times 10^{-24}\ kg\ m\ s^{-1}; \{***\}. \qquad (2.7b)$$

$$(2010)\ \hbar/a_o = 1.992\ 851\ 741\ 13 \times 10^{-24}\ kg\ m\ s^{-1}; \{****\}. \qquad (2.7b)$$

2006 (**) CODATA atomic unit of momentum = $1.992\ 851\ 565 \times 10^{-24}$ kg m s^{-1}. In atomic unit of momentum, the CODATA constant and our calculations agree to eight decimal places. We exhibit the same precision as the 2006 and 2010 CODATA recommended values. Please observe, when the atomic unit of momentum, $\hbar/a_o$, equation (2.7a), is multiplied by $(4\pi_{qed})$ quantization constant, the product is the natural unit of momentum p, equation (2.6a).

Compton wavelength:

$$\lambda_c = 1/[2\ R_\infty (4\pi_{qed})^2]; \{\$\}. \qquad (2.8a)$$

$(2006)\ \lambda_c = 2.426\ 310\ 217\ 51\ \underline{67}\ \text{x}\ 10^{-12}\ \text{m}\ \text{c}^{-1}\ \{*\}.$ (2.8b)

$(2006)\ \lambda_c = 2.426\ 310\ 217\ 5\ (33)\ \text{x}\ 10^{-12}\ \text{m}\ \text{c}^{-1}\ \{**\}.$ (2.8b)

$(2010)\ \lambda_c = 2.426\ 310\ 238\ 9\ (16)\ \text{x}\ 10^{-12}\ \text{m}\ \text{c}^{-1}\ \{***\}.$ 2.8b)

$(2010)\ \lambda_c = 2.426\ 310\ 238\ 96\ \text{x}\ 10^{-12}\ \text{m}\ \text{c}^{-1}\ \{****\}.$ (2.8b)

2006 (**) CODATA value: $\lambda_c = 2.426\ 310\ 2175\ \text{x}\ 10^{-12}\ (33)$ m. Here, the CODATA value and our calculations agree to ten decimal places. We exhibit higher precision that is one order of magnitude better than that of the 2006 and 2010 CODATA recommended constants.

Natural unit of conductance:

$$G_r = (\varepsilon_0/\mu_0)^{1/2}.$$ (2.9a)

$$G_r = 2.654\ 418\ 729\ 438\ 0724\ \text{x}\ 10^{-3}\ \text{S}.$$ (2.9b)

CODATA do not have this constant. We use the natural unit of conductance to generate conductance quantum. Our precision here is to fifteen decimal places, i.e., standard uncertainty and relative standard uncertainty is exact.

Conductance quantum:

$$G_o = G_r/(\pi_{qed});\ \{\$\}.$$ (2.10a)

$(2006)\ G_o = 7.748\ 091\ 700\ 44\underline{63}\ \text{x}\ 10^{-5}\ \text{S};\ \{*\}.$ (2.10b)

$(2006)\ G_o = 7.748\ 091\ 700\ 4\ \text{x}\ 10^{-5}\ \text{S};\ \{**\}.$ (2.10b)

$(2010)\ G_o = 7.748\ 091\ 734\ 6\ (25)\ \text{x}\ 10^{-5}\ \text{S};\ \{***\}.$ (2.10b)

$(2010)\ G_o = 7.748\ 091\ 734\ 630\ 9\ \text{x}\ 10^{-5}\ \text{S};\ \{****\}.$ (2.10b)

2006 (**) CODATA value: $G_o = 2e^2/h = 7.748\ 091\ 7004\ \text{x}\ 10^{-5}\ (53)$ S. In conductance quantum, the CODATA value and our calculations agree to ten decimal places. We show higher precision by a factor of ten than the 2006 and 2010 CODATA recommended values. Notice, when the conductance quantum, G_o, equation (2.10a), is multiplied by (π_{qed}) quantization constant, the product is the natural unit of conductance, G_r, equation (2.9a).

Natural unit of resistance (characteristic impedance of vacuum):

$$Z_o = (\mu_o/\varepsilon_o)^{1/2}. \tag{2.11a}$$

$$(2006)\ Z_o = 376.730\ 313\ 461\ 7707 \cdots \Omega;\ \{*\}. \tag{2.11b}$$

$$(2006)\ Z_o = 376.730\ 313\ 461 \cdots \Omega;\ \{**\}. \tag{2.11b}$$

$$(2010)\ Z_o = 376.730\ 313\ 461 \cdots \Omega;\ \{***\}. \tag{2.11b}$$

$$(2010)\ Z_o = 376.730\ 313\ 461\ 7707 \cdots \Omega;\ \{****\}. \tag{2.11b}$$

2006 (**) CODATA value: $Z_o = 376.730\ 313\ 461 \cdots \Omega$. Here, the CODATA value and our calculations agree to eleven decimal places. We exhibit higher precision that are three orders of magnitude better than that of the 2006 and 2010 CODATA recommended values, i.e., standard uncertainty and relative standard uncertainty is exact.

Inverse of conductance quantum:

$$G_o^{-1} = Z_o(\pi_{qed});\ \{\$\}. \tag{2.12a}$$

$$(2006)\ G_o^{-1} = 12\ 906.403\ 7786\ \underline{54}\ \Omega;\ \{*\}. \tag{2.12b}$$

$$(2006)\ G_o^{-1} = 12\ 906.403\ 7787\ (88)\ \Omega;\ \{**\}. \tag{2.12b}$$

$$(2010)\ G_o^{-1} = 12\ 906.403\ 7217\ (42)\ \Omega;\ \{***\}. \tag{2.12b}$$

$$(2010)\ G_o^{-1} = 12\ 906.403\ 7217\ 11\ \Omega;\ \{****\}. \tag{2.12b}$$

2006 (**) CODATA value: $G_o^{-1} = h/2e^2 = 12\ 906.403\ 7787\ (88)\ \Omega$. Here, the CODATA value and our calculations agree to ten decimal places. We show higher precision by a factor of ten than the 2006 and 2010 CODATA recommended values. Please note, according to '**1**' in the principle of the one-and-the-many, resistance and conductance are inverse quantities. Therefore, to obtain the natural unit of resistance Z_o, equation (2.11a), the inverse of conductance quantum, G_o, equation (2.12a), is divided by (π_{qed}) quantization constant.

Natural unit of electric potential (electron mass energy equivalent):

$$V = R_\infty e(4\pi_{qed})^3/\varepsilon_0; \{\$\}. \tag{2.13a}$$

$$(2006)\ V = 510\ 998.909\ 5\underline{09}\ V;\ \{^*\}. \tag{2.13b}$$

$$(2006)\ V = 510\ 998.910\ (13)\ V;\ \{^{**}\}. \tag{2.13b}$$

$$(2010)\ V = 510\ 998.928\ V;\ \{^{***}\}. \tag{2.13b}$$

$$(2010)\ V = 510\ 998.927\ 617\ V;\ \{^{****}\}. \tag{2.13b}$$

2006 (**) CODATA value: $V = 510\ 998.910\ (13)$ V. Here, the CODATA value and our calculations agree to six decimal places. We display higher precision that is one order of magnitude better than that of the 2006 and 2010 CODATA recommended constants.

Atomic unit of electric potential:

$$E_h/e = V/(4\pi_{qed})^2; \{\$\}. \tag{2.14a}$$

$$(2006)\ E_h/e = 27.211\ 383\ 853\ \underline{56}\ V;\ \{^*\}. \tag{2.14b}$$

$$(2006)\ E_h/e = 27.211\ 383\ 86\ (68)\ V;\ \{^{**}\}. \tag{2.14b}$$

$$(2010)\ E_h/e = 27.211\ 385\ 05\ (60)\ V;\ \{^{***}\}. \tag{2.14b}$$

$$(2010)\ E_h/e = 27.211\ 385\ 0579\ V;\ \{^{****}\}. \tag{2.14b}$$

2006 (**) CODATA value: $E_h/e = 27.211\ 383\ 86\ (68)$ V. Here, the CODATA value and our calculations agree to eight decimal places. We exhibit higher precision by a factor of ten than the 2006 and 2010 CODATA recommended values. Notice, just as with energy quantization, when the atomic unit of electric potential, E_h/e, equation (2.14a), is multiplied by $(4\pi_{qed})^2$ quantization constant, the product is the natural unit of electric potential V, equation (2.13a).

Josephson constant:

$$K_J = 1/(\pi_{qed})\mu_o ve; \ \{\$\}. \tag{2.15a}$$

$$(2006) \ K_J = 483 \ 597. \ 890 \ 951 \ \underline{09} \ \text{x} \ 10^9 \ \text{Hz} \ \text{V}^{-1}; \ \{*\}. \tag{2.15b}$$

$$(2006) \ K_J = 483 \ 597. \ 891 \ (12) \ \text{x} \ 10^9 \ \text{Hz} \ \text{V}^{-1}; \ \{**\}. \tag{2.15b}$$

$$(2010) \ K_J = 483 \ 597. \ 870 \ (11) \ \text{x} \ 10^9 \ \text{Hz} \ \text{V}^{-1}; \ \{***\}. \tag{2.15b}$$

$$(2010) \ K_J = 483 \ 597. \ 869 \ 541 \ 137 \ \text{x} \ 10^9 \ \text{Hz} \ \text{V}^{-1}; \ \{****\}. \tag{2.15b}$$

2006 (**) CODATA value: $K_J = 2e/h$ = 483 597. 891 (12) x 10^9 Hz V^{-1}. In the Josephson constant, the CODATA value and our calculations agree to seven decimal places. We show higher precision that is one order of magnitude better than that of the 2006 and 2010 CODATA recommended values.

von Klitzing constant:

$$R_k = (2\pi_{qed})\mu_o v; \ \{\$\}. \tag{2.16a}$$

$$(2006) \ R_k = 25 \ 812.807 \ 557 \ 3\underline{08} \ \Omega; \ \{*\}. \tag{2.16b}$$

$$(2006) \ R_k = 25 \ 812.807 \ 557 \ (18) \ \Omega; \ \{**\}. \tag{2.16b}$$

$$(2010) \ R_k = 25 \ 812.807 \ 4434 \ (84) \ \Omega; \ \{***\}. \tag{2.16b}$$

$$(2010) \ R_k = 25 \ 812.807 \ 4434 \ 23 \ \Omega; \ \{****\}. \tag{2.16b}$$

2006 (**) CODATA value: $R_k = h/e^2$ = 25 812.807 557 (18) Ω. In the von Klitzing constant, the CODATA value and our calculations agree to ten decimal places. We show higher precision that is one order of magnitude better than that of the 2006 and 2010 CODATA recommended values.

Quantum of circulation:

$$h/2m_e = v/4 \ R_\infty (4\pi_{qed})^2 = vS(\pi_{qed}); \ \{\$\}. \tag{2.17a}$$

$$(2006) \ h/2m_e = 3.636 \ 947 \ 519 \ 89\underline{22} \ \text{x} \ 10^{-4} \ \text{m}^2 \ \text{s}^{-1}; \ \{*\}. \tag{2.17b}$$

$$(2006) \ h/2m_e = 3.636 \ 947 \ 519 \ 9 \ (50) \ \text{x} \ 10^{-4} \ \text{m}^2 \ \text{s}^{-1}; \ \{**\}. \tag{2.17b}$$

(2010) $h/2m_e = 3.636\ 947\ 552\ 0\ (24) \times 10^{-4}\ m^2\ s^{-1}$; {***}.   (2.17b)

(2010) $h/2m_e = 3.636\ 947\ 552\ 037 \times 10^{-4}\ m^2\ s^{-1}$; {****}.   (2.17b)

2006 (**) CODATA value: $h/2m_e = 3.636\ 947\ 5199\ (50) \times 10^{-4}\ m^2\ s^{-1}$. In quantum of circulation, the CODATA value and our calculations agree to nine decimal places. We show higher precision that is an order of magnitude better than that of the 2006 and 2010 CODATA recommended value.

Magnetic flux quantum:

$$\Phi_o = \phi\pi_{qed};\ \{\$\}.\tag{2.18a}$$

(2006) $\Phi_o = 2.067\ 833\ 666\ \underline{59} \times 10^{-15}\ Wb$; {*}.   (2.18b)

(2006) $\Phi_o = 2.067\ 833\ 667\ (52 \times 10^{-15}\ Wb$; {**}.   (2.18b)

(2010) $\Phi_o = 2.067\ 833\ 758\ (46) \times 10^{-15}\ Wb$; {***}.   (2.18b)

(2010) $\Phi_o = 2.067\ 833\ 758\ 14 \times 10^{-15}\ Wb$; {****}.   (2.18b)

2006 (**) CODATA value: $\Phi_o = h/2e = 2.067\ 833\ 667\ (52) \times 10^{-15}\ Wb$. In magnetic flux quantum, the CODATA value and our calculations agree to eight decimal places. For 2010 natural unit of magnetic flux (flux of magnetic induction), ϕ, ($6.035\ 884\ 795\ 5 \times 10^{-17}\ Wb$ (****)) see equation (3.16).

Atomic unit of velocity:

$$a_oE_h/\hbar = v/(4\pi_{qed});\ \{\$\}.\tag{2.19a}$$

(2006) $a_oE_h/\hbar = 2.187\ 691\ 254\ 13\underline{94} \times 10^6\ m\ s^{-1}$; {*}.   (2.19b)

(2006) $a_oE_h/\hbar = 2.187\ 691\ 254\ 1\ (15) \times 10^6\ m\ s^{-1}$; {**}.   (2.19b)

(2010) $a_oE_h/\hbar = 2.187\ 691\ 263\ 79\ (71) \times 10^6\ m\ s^{-1}$; {***}.   (2.19b)

(2010) $a_oE_h/\hbar = 2.187\ 691\ 263\ 7915 \times 10^6\ m\ s^{-1}$; {****}.   (2.19b)

(2006) CODATA value: $a_o E_h / \hbar$ = 2.187 691 2541 (15) x 10^6 m s^{-1}. In atomic unit of velocity, the CODATA value and our calculations agree to eleven decimal places. We show higher precision by a factor of ten than the 2006 and 2010 CODATA recommended values. For natural unit of velocity, v, see equation (3.28).

To get to the heart of the matter, at the *quantum* level, equations (2.10a), (2.12a), (2.17a), and (2.18a), (π_{qed}) quantization takes place; at the *atomic unit* level, equations (2.7a) and (2.19), ($4\pi_{qed}$), or equations (2.4a), (2.5a), and (2.14a), ($4\pi_{qed}$)^2 quantization transpires.

In paper 3 we predict fifty new fundamental physical constants. In paper 4 we utilize papers 2 and 3 constants to predict the order of the quantities and the natural spectrum of physical quantities and to give form to the LSPR.

References

[1] Penrose R (2004) *The Road to Reality: A complete guide to the Laws of the Universe* (Jonathan Cape, London).

[2] Elitzur A, Dolev S, Kolenda N, Eds (2005) *Quo Vadis Quantum Mechanics?* (Springer, Berlin).

[3] Feynman R P, Leighton R B, Sands M (1965) *The Feynman Lectures on Physics*. Vol 3 (Addison-Wesley, Reading, PA).

[4] Weinberg S (2000) *The Quantum Theory of Fields*. Vol III, Supersymmetry (Cambridge University Press, Cambridge).

[5] Bohm D, Hiley B J (1959) Non-Locality and Locality in the Stochastic Interpretation of Quantum Mechanics. *Physics Reports* **172**, 93-122.

[6] Bell J S (1964) On the Einstein Podolsky Rosen Paradox. *Physics* 1, 195-200.

[7] Bedrij O (1993) Scale Invariance, Unifying Principle, Order and Sequence of Physical Quantities and Fundamental Constants. *Proc Inst Math Natl Acad Sci Ukraine*; also, (1994) *Ukrainian Mathematical Journal,* tr of the Proc Inst Math Natl Acad Sci Ukraine (Allerton Press, New York), pp 67-73.

[8] Mohr P J, Taylor B N (2005) CODATA Recommended Values of the Fundamental Physical Constants: 2002. *Rev Mod Phys* vol 77(1), pp 1-107.

[9] Bohr N (1934) *Atomic Theory and the Description of Nature* (Cambridge University Press, Cambridge).

[10] Hantzsche E (1990) Elementary Constants of Nature. *Annalen der Physik* 7. Folge, Band 47. Heft 5, S. 401-412. (Akademie der Wissenschaften der DDR, Zentralinstitut Fur Electronenphysik, Berlin.)

[11] Stroke H H, ed (1995) *The Physical Review: The First Hundred Years* (Amer Inst of Phys, Woodbury, NY).

[12] Schrödinger E (1928) *Collected Papers on Wave Mechanics* (Blackie).

[13] Einstein A (1956) *The Meaning of Relativity* (Princeton University Press, Princeton, NJ).

[14] Heisenberg W (1930) *The Principles of the Quantum Theory* (University of Chicago Press).

3

Prediction of the New Fundamental Physical Constants

Thus, the task is, not so much to see what no one has yet seen; but to think what nobody has yet thought, about that which everybody sees.

—Erwin Schrödinger,
1887-1961

Abstract

In paper 3, through '**1**', the laws of physics, four very high-precision fundamental physical constants, and our suggested quantization constant, we predict fifty new physical constants. At this point we draw on 2006 CODATA recommended values. By means of these predictions, dimensional uniformity of physical equations, constants, and symmetries of electromagnetism with gravitation are determined. We predict the fundamental constants and relationships for the natural unit of gravitational potential and field strength and the natural unit of the universal gravitational constant in QED.

In addition, we predict five new gravitational quantities, their physical relationships, and the values of their fundamental constants in QED: the gravitational constant, the natural unit of gravitational flux density, the natural unit of flux of the gravitational field, the gravitational vector potential, and the natural unit of gravitance.

1 Introduction

Utilizing equations (1.9b)-(1.13b) 2006 constants and the laws of physics, we have calculated fifty new constants in QED that CODATA do not provide. These constants are required to produce the LSPR. [1] They can also be effective in calculating physical processes. "For the present," says S. Weinberg, "the calculations of physical processes at the Planck energy is simply beyond our reach." [2] The LSPR is a powerful research tool used for prediction and verification of the laws of physics and the fundamental physical constants beyond the Planck scale. The constants are the following:

Natural unit of magnetic potential, equation (3.1)
Natural unit of electric power, equation (3.2)
Natural unit of time, equation (3.3b)
Natural unit of length, equation (3.4b)
Natural unit of area, equation (3.5b)
Natural unit of volume, equation (3.6b)
Natural unit of capacitance, equation (3.7b)
Natural unit of self-inductance, equation (3.8b)
Characteristic impedance of vacuum, equation (3.9)
Natural unit of electric field strength, equation (3.11)
Natural unit of magnetic field strength, equation (3.12)
Natural unit of electric flux density, equation (3.13)
Natural unit of magnetic flux density, equation (3.14)
Natural unit of electric flux, equation (3.15)
Natural unit of magnetic flux, equation (3.16)
Magnetic flux quantum, equation (3.17)
Natural unit of flux of the electric field, equation (3.18)
Natural unit of flux of the magnetic field, equation (3.19)
Natural unit of force, equation (3.20b)
Natural unit of pressure, equation (3.21b)
Natural unit of action, equation (3.22b)
Natural unit of volume flow rate, equation (3.23)
Natural unit of coefficient of viscosity, equation (3.24b)
Natural unit of charge density, equation (3.25)
Natural unit of mass density, equation (3.26)
Natural unit of mass flow rate, equation (3.27b)
Natural unit of velocity, equation (3.28)
Atomic unit of velocity, equation (3.29)

Natural unit of angular velocity, equation (3.30b)
Natural unit of frequency, equation (3.31)
Natural unit of period of harmonic motion, equation (3.32)
Natural unit of gravitational flux density, equation (3.33b)
Natural unit of linear mass density, equation (3.34b)
Natural unit of gravitational field strength, equation (3.35b)
Natural unit of gravitational potential, equation (3.36b)
Natural unit of the universal gravitational constant, equation (3.37b)
Penetrability of free space (gravitational constant), equation (3.38b)
Natural unit of gravitance, equation (3.39b)
Natural unit of electric dipole moment, equation (3.40b)
Natural unit of magnetic dipole moment, equation (3.41b)
Natural unit of electric vector potential, equation (3.42)
Natural unit of magnetic vector potential, equation (3.43)
Natural unit of electric conductivity, equation (3.44)
Natural unit of electric resistivity, equation (3.45)
Natural unit of voltage density, equation (3.46)
Natural unit of current density, equation (3.47)
Natural unit of compressibility, equation (3.48b)
Natural unit of particle fluence, equation (3.49b)
Natural unit of moment of inertia, equation (3.50)
Natural unit of angular acceleration, equation (3.51b)
Natural unit of angle of rotation, equation (3.52b)
Natural unit of (linear) time, equation (3.53b)
Natural unit of circular wave number, equation (3.54b)
Natural unit of intensity, equation (3.55)
Natural unit of angular momentum, equation (3.56b)
Rydberg constant, equation (3.57b)
Natural unit of flux of the gravitational field, equation (3.58b)

2 Suggested Quantization Constant

In Euclidean plane geometry, π is defined as the ratio of a circle's circumference, C_π to its diameter, d_π: $\pi = C_\pi/d_\pi$. In the dynamic n rotation π of an electron, we incorporate in our calculations the rotation group of the electron's underlying gyratory motion with respect to the reference frame of the cosmos. For example, in reference to the rest frame '1', the electron velocity, in the orbital $\lambda = 1$, is $\sim 2.42 \times 10^6$ m/s, the Local Group and the Milky Way galaxy velocity is $\sim 6 \times 10^5$ m/s, while the solar system velocity is ~ 3.59 ± 1.80 x 10^5 m/s. [3] In addition, the orbits the electrons, planets, and galaxies travel on are also themselves undergoing universal expansion. If we consider frequency, f, and angular velocity, ω, as independent quantities, one can obtain from equations (3.30a)-(3.31), (3.52a), and (3.56a) the following relationships:

$$\omega/f = eBh/Em_e = \lambda_c/S \qquad (3.0a)$$

The suggested QED quantization constant, π_{qed}, is calculated as follows: [1, 4]

$$(2006)\ \pi_{qed} = \omega/2f = 34.258\ 999\ 919\ 74\ \{\$, \#\}, \qquad (3.0b)$$

$$(2010)\ \pi_{qed} = \omega/2f = 34.258\ 999\ 768\ 6\ \{\$, \#\}, \qquad (3.0b)$$

and with equation (3.52a), we get the following:

$$(2006)\ \omega/f = \theta = 2\pi_{qed} = 1/2\alpha = 68.517\ 999\ 8395; \{\$, \#\}, \qquad (3.0c)$$

$$(2010)\ \omega/f = \theta = 2\pi_{qed} = 1/2\alpha = 68.517\ 999\ 5372; \{\$, \#\}, \qquad (3.0c)$$

where α is the fine-structure constant, and θ is the natural unit of the angle of rotation, equation (3.52b). As can be seen, the $\omega/f \neq 2\pi$ as it is accepted in QM and the Heisenberg uncertainty principle equations employing the mathematical constant 2π value [5] in the rotation group of the spin space.

The P. A. M. Dirac relationship describing motion of an elementary particle with spin ½ (electron or proton) does not consider dimensional uniformity of physical equations in '1'. Yet it is an inseparable part of modern mathematical and theoretical physics. [6, 7] In physical terms, this corresponds to the fact that there is no measurement protocol that can specify a basis of the spin space. We suggest that the Dirac's constant $h/2\pi = \hbar$ will vary with the velocity in the electron, proton, and neutron. [1, 8, 9] Also, the universal gravitation constant G over h-bar c, $G/\hbar c$, will alter by way of the rotation group in the given number of dimensions [10, 11, 12] as is shown in equations (3.37a)-(3.37b). As a result, we suggest that they are ill defined in the existing form for QED [13] and Einstein's general relativity equations using the "universal gravitation constant" with other fundamental constants. [1, 4, 5, 10]

The number 2π dominates the usual laws of physics describing fundamental principles of the universe. These principles, such as the Heisenberg indeterminacy relations, the quantization of action, Einstein's field equation of general relativity, Planck energy, etc., are dynamic. We suggest that in quantum production and measurements in QED, $(2\pi_{qed}) = (1/2\alpha)$ be utilized instead 2π.

3 Prediction Results

In equations (3.1)-(3.61), we are deriving constants without the 2π or the π_{qed}. Equations (3.1)-(3.61) are based on the laws of physics with equations (1.9b)-(1.13b) *verified* QED constants of equations (2.1a)-(2.19), where the π_{qed} was utilized. By way of illustration, the natural unit of magnetic potential i, equation (3.1), equals the natural unit of electric potential V, equation (2.13b), times the natural unit of conductance G_r, equation (2.9b). Prediction results of the new constants are determined as follows:

Natural unit of magnetic potential (electric current):

$$i = G_r V = 1\ 356.405\ 076\ \underline{13}\ \text{A}; \{\#\}. \tag{3.1}$$

Natural unit of electric power:

$$P = iV = 693\ 121\ 514.7\underline{53}\ \text{J s}^{-1}; \{\#\}. \tag{3.2}$$

Natural unit of (angular) time:

$$t = e/i = E/P = 1/R_\infty v(4\pi_{qed})^3; \{\$\}. \tag{3.3a}$$

$$t = 1.181\ 193\ 225\ 5\underline{346} \times 10^{-22}\ \text{s r}^{-1}; \{\#\}, \tag{3.3b}$$

CODATA value: [4] $t_c = \hbar/m_e c^2 = 1.288\ 088\ 6570 \times 10^{-21}$ (18) s. Notice, the natural unit of time, $t = e/i$, is determined via the elementary charge, e, equation (1.13b), and magnetic potential, i, equation (3.1). The CODATA natural unit of time is derived by the $(h/2\pi)/m_e c^2$ relationship. When the 2π in the CODATA equation is corrected to the QED value of $(2\pi_{qed})$, then both constants will be $t = 1.1812 \times 10^{-22}$. Also, please see equation (3.10).

The unit radian, r, has historically been designated as a supplementary unit. In 1980, the International Committee for Weights and Measures determined that angle and solid angle are to be regarded as dimensionless derived quantities. According to the committee, the unit radian and steradian are equivalent to the number 1 and may be omitted for derived units. As the natural unit of the angle of rotation, equation (3.52b), expressed in radians that correspond to one revolution, or a cycle of 360°, is a physical quantity, we suggest that it should be included in the definition of quantities.

Natural unit of (angular) length:

$$S = vt = m_e v/eB = 1/R_\infty (4\pi_{qed})^3 = \lambda_c/2\pi_{qed}; \{\$\}. \tag{3.4a}$$

$$S = 3.541\ 128\ 204\ 55\underline{97} \times 10^{-14}\ \text{m r}^{-1}; \{\#\}. \tag{3.4b}$$

CODATA value: $S_c = \hbar/m_e c = 386.159\ 264\ 59 \times 10^{-15}$ (53) m. The CODATA natural unit of length is derived via the $(h/2\pi)/m_e c$ relationship. When the 2π in the CODATA equation is corrected to the QED value of $1/2\alpha$, then both constants are $S = 3.5411 \times 10^{-14}$. Also, when the classical electron radius (r_e) is multiplied by 4π, we obtain $S = (4\pi)(r_e)$, where $r_e = \alpha^3/4pR_\infty = 2.817\ 940\ 2894 \times 10^{-15}$ m.

Natural unit of area:

$$A = S^2 = 1/R_\infty^{\ 2}(4\pi_{qed})^6; \{\$\}. \tag{3.5a}$$

$$A = 1.253\ 958\ 896\ 11\underline{28} \times 10^{-27}\ \text{m}^2\ \text{r}^{-2}; \{\#\}. \tag{3.5b}$$

Natural unit of volume:

$$V_o = S^3 = m_e/d = 1/R_\infty^{\ 3}(4\pi_{qed})^9; \{\$\}. \tag{3.6a}$$

$$V_o = 4.440\ 429\ 214\ 38\underline{38} \times 10^{-41}\ \text{m}^3\ \text{r}^{-3}; \{\#\}. \tag{3.6b}$$

For natural unit of (mass) density, d, see equation (3.26). To compare, the proton volume is $V_{op} = 1.075827 \times 10^{-36}\ \text{m}^3\ \text{r}^{-3}$, while the neutron volume is $V_{on} = 1.077819 \times 10^{-36}\ \text{m}^3\ \text{r}^{-3}$ [8].

Natural unit of capacitance:

$$C = e/V = S\varepsilon_o; \{\$\}. \tag{3.7a}$$

$$C = 3.135\ 381\ 420\ 94\underline{44} \times 10^{-25}\ \text{F}; \{\#\}. \tag{3.7b}$$

Natural unit of self-inductance:

$$L = S\mu_o; \{\$\}. \tag{3.8a}$$

$$L = 4.449\ 912\ 941\ 1458 \times 10^{-20}\,\text{H}; \{\#\}. \tag{3.8b}$$

In equation (2.11a), we have calculated characteristic impedance of vacuum through the permeability and permittivity of free space. Here, in equation (3.9), we verify the characteristic impedance of a vacuum through capacitance and self-inductance:

$$Z_o = (L/C)^{1/2} = 376.730\ 313\ 461\ 7707 \cdots \Omega. \tag{3.9}$$

Equally, via capacitance, self-inductance, and characteristic impedance of a vacuum, we verify the natural unit of *time constant*, equation (3.3b):

$$t = (LC)^{1/2} = Z_o C = L/Z_o = 1.181\ 193\ 225\ 5\underline{346} \times 10^{-22}\,\text{s r}^{-1}. \tag{3.10}$$

Natural unit of electric field strength:

$$E_s = V/S = 1.443\ 039\ 844\ \underline{91} \times 10^{19}\,\text{V m}^{-1}; \{\#\}. \tag{3.11}$$

Natural unit of magnetic field strength:

$$H = i/S = 3.830\ 431\ 991\ \underline{64} \times 10^{16}\,\text{A m}^{-1}; \{\#\}. \tag{3.12}$$

Natural unit of electric flux density (surface charge):

$$D = H/v = E_s \varepsilon_o = e/A = 127\ 769\ 458.151\,\text{C m}^{-2}; \{\#\}. \tag{3.13}$$

Natural unit of magnetic flux density (magnetic induction):

$$B = E_s/v = H\mu_o = 4.813\ 462\ 802\ \underline{01} \times 10^{10}\,\text{T}; \{\#\}. \tag{3.14}$$

In equation (1.13b), we utilized the 2006 CODATA recommended value for the elementary charge of the electron. Here, in equation (3.15), we verify the elementary charge with other constants.

Natural unit of electric flux (atomic unit of charge):

$$e = CV = it = DA = E/V = 1.602\ 176\ 487 \times 10 \times 10^{-19} \text{ C; } \{\#\}. \quad (3.15)$$

Natural unit of magnetic flux (flux of magnetic induction):

$$\phi = Li = Vt = BA = E/i = 6.035\ 884\ 501\ \underline{69} \times 10^{-17} \text{ Wb; } \{\#\}. \quad (3.16)$$

Magnetic flux quantum:

$$\Phi_{o} = \phi\pi_{qed} = 2.067\ 833\ 666\ \underline{59} \times 10^{-15} \text{ Wb; } \{\$, *\}. \quad (3.17)$$

CODATA value: $\Phi_{o} = h/2e = 2.067\ 833\ 667 \times 10^{-15}$ (52) Wb. In magnetic flux quantum, the CODATA value and our calculations agree to eight decimal places.
Natural unit of flux of the electric field:

$$\Phi_{E} = E_{s}A = e/\varepsilon_{o} = 1.809\ 512\ 650\ \underline{96} \times 10^{-8} \text{ V-m; } \{\#\}. \quad (3.18)$$

Natural unit of flux of the magnetic field:

$$\Phi_{H} = HA = \phi/\mu_{o} = 4.803\ 204\ 271\ \underline{87} \times 10^{-11} \text{ A-m; } \{\#\}. \quad (3.19)$$

Natural unit of force:

$$F = eE_{s} = \phi H = P/v = m_{e}v/t = E/S. \quad (3.20a)$$

$$F = 2.312\ 004\ 509\ \underline{31} \text{ N; } \{\#\}. \quad (3.20b)$$

Natural unit of pressure (energy density):

$$P_{r} = F/A = BH = DE_{s} = E/V_{o}. \quad (3.21a)$$

$$P_{r} = 1.843\ 764\ 190\ \underline{73} \times 10^{27} \text{ Pa; } \{\#\}. \quad (3.21b)$$

Natural unit of action (atomic unit of action):

$$\hbar_{e} = h/2\pi_{qed} = e\phi = Et = pS; \{\$\}. \quad (3.22a)$$

$$\hbar_{e} = 9.670\ 552\ 226\ \underline{84} \times 10^{-36} \text{ J s; } \{\$, \#\}. \quad (3.22b)$$

CODATA value: $\hbar c = h/2\pi = 1.054\ 571\ 628 \times 10^{-34}$ (53) J s. The CODATA natural unit of action is derived from the $h/2\pi$ relationship. When the 2π in the CODATA equation is corrected to the QED value of $1/2\alpha$, then both constants are the same. Also, see equation (3.56b).

Natural unit of volume flow rate (volume flux):

$$Q = V_0/t = Av = 3.759\ 274\ 196\ 9\underline{664} \times 10^{-19} \text{ m}^3 \text{ s}^{-1}; \{\#\}. \qquad (3.23)$$

Natural unit of coefficient of viscosity:

$$\eta = E/Q = e\phi/V_0 = tP_r = \hbar_e/V_0; \{\$\}. \qquad (3.24\text{a})$$

$$\eta = 217\ 784.177\ 1\underline{58} \text{ Pa s}; \{\#\}. \qquad (3.24\text{b})$$

Natural unit of charge density:

$$\rho_c = e/V_0 = 3.608\ 156\ 801\ \underline{17} \times 10^{21} \text{ C m}^{-3}; \{\#\}. \qquad (3.25)$$

Natural unit of (mass) density:

$$d = m_e/V_0 = 2.051\ 464\ 330\ \underline{17} \times 10^{10} \text{ kg m}^{-3}; \{\#\}. \qquad (3.26)$$

Natural unit of mass flow rate:

$$Q_m = m_e/t = Qd = eB = fD \qquad (3.27\text{a})$$

$$Q_m = 7.712\ 016\ 922\ \underline{43} \times 10^{-9} \text{ kg s}^{-1}; \{\#\}. \qquad (3.27\text{b})$$

Natural unit of velocity: v, where in QED v = c,

$$v = H/D = E_s/B = P/F = 299\ 792\ 458 \text{ m s}^{-1}. \qquad (3.28)$$

CODATA value: 299 792 458 m s⁻¹.

Atomic unit of velocity:

$$v/4\pi_{qed} = 2.187\ 691\ 254\ 13\underline{94} \times 10^6 \text{ m s}^{-1}; \{\$, *\}. \qquad (3.29)$$

CODATA value: $2.187\ 691\ 2541 \times 10^6\ (15)$ m s^{-1}. In atomic unit of velocity, the CODATA value and our calculations agree to ten decimal places. We show higher precision that is one order of magnitude better than that of the 2006 CODATA recommended value.

Natural unit of angular velocity (the cyclotron frequency):

$$\omega = eB/m_e = v/S = E/\hbar_e; \{\$\}. \tag{3.30a}$$

$$\omega = 8.466\ 015\ 368\ 03\underline{92} \times 10^{21}\ \text{r s}^{-1}; \{\#\}. \tag{3.30b}$$

Natural unit of frequency:

$$f = E/h = v/\lambda_c = \omega/2\pi_{qed} = 1.235\ 589\ 974\ 58\underline{63} \times 10^{20}\ \text{Hz}; \{\$, \#\}. \tag{3.31}$$

Natural unit of period of harmonic motion (cyclic or periodic time):

$$T = 1/f = \lambda_c/v = 2\pi_{qed}/\omega = 8.093\ 299\ 723\ 76\underline{00} \times 10^{-21}\ \text{Hz}^{-1}; \{\$, \#\}. \tag{3.32}$$

Comparable to the natural unit of electric flux density $D = e/A$, equation (3.13), and the natural unit of magnetic flux density $B = \phi/A$, equation (3.14), we propose a new physical quantity: the natural unit of gravitational flux density M.

Natural unit of gravitational (surface) flux density:

$$M = m_e/A = \eta/v = gz_o; \{\$\}. \tag{3.33a}$$

$$M = 7.264\ 498\ 200\ \underline{23} \times 10^{-4}\ \text{kg m}^{-2}; \{\$, \#\}. \tag{3.33b}$$

Natural unit of linear mass density (gravitational vector potential):

$$\mu = m_e/S = V_g z_o; \{\$, \#\}. \tag{3.34a}$$

$$\mu = 2.572\ 451\ 946\ \underline{88} \times 10^{-17}\ \text{kg m}^{-1}; \{\#\}. \tag{3.34b}$$

Natural unit of gravitational field strength:

$$g = F/m_e = v\omega = P_r/M; \{\$\}. \tag{3.35a}$$

$$g = 2.538\ 047\ 556\ 6503 \times 10^{30}\ \text{m s}^{-2}; \{\#\}. \tag{3.35b}$$

Natural unit of gravitational potential (linear stopping power):

$$V_g = gS = 1/\varepsilon_0\mu_0 = eV/m_e; \{\$\}. \qquad (3.36a)$$

$$V_g = 8.987\ 551\ 787\ 368\ 178\ \cdots \times 10^{16}\ J\ kg^{-1}; \{\#\}. \qquad (3.36b)$$

Natural unit of the universal gravitation constant (in QED):

$$G_e = g/M = V_g/\mu = SV_g/m_e = 1/dLC; \{\$\}. \qquad (3.37a)$$

$$G_e = 3.493\ 768\ 580\ \underline{69} \times 10^{33}\ m^3\ kg^{-1}\ s^{-2}; \{\#\}. \qquad (3.37b)$$

According to Mohr and Taylor, [4] at present "there is no known quantitative theoretical relationship between the universal gravitation constant G and other fundamental constants." In equations (3.37a)-(3.39a), we show this relationship.

Analogous to the permeability of free space (magnetic constant), $\mu_0 = L/S$, equation (1.9a), and the permittivity of free space (electric constant), $\varepsilon_0 = C/S$, equation (1.10a), we suggest a new physical quantity: penetrability of free space (gravitational constant), z_o.

Penetrability of free space (gravitational constant):

$$z_o = \Omega/S = 1/G_e = M/g = \Omega R_\infty (4\pi_{qed})^3; \{\$\}. \qquad (3.38a)$$

$$z_0 = 2.862\ 238\ 802\ \underline{89} \times 10^{-34}\ kg\ s^2\ m^{-3}; \{\#\}. \qquad (3.38b)$$

Comparable to the natural unit of capacitance $C = e/V$, equation (3.7a), in the electric field and the natural unit of self-inductance $L = \phi/i$, equation (3.16), in the magnetic field, we propose a new quantity: natural unit of gravitance in the gravitational field Ω.

Natural unit of gravitance:

$$\Omega = m_e/V_g = Sz_0 = MLC; \{\$\}. \qquad (3.39a)$$

$$\Omega = 1.013\ 555\ 455\ \underline{31} \times 10^{-47}\ kg^2\ J^{-1}; \{\#\}. \qquad (3.39b)$$

Additional gravitational quantities and comparison of electron, proton, neutron, and earth calculations and constants can be found in [10], table 4.

Natural unit of electric dipole moment:

$$p_d = eS = DV_o = E/E_s. \qquad (3.40a)$$

$$p_d = 5.673\ 512\ 346\ \underline{79} \times 10^{-33}\ C\ m;\ \{\#\}. \qquad (3.40b)$$

Natural unit of magnetic dipole moment:

$$j = \phi S = BV_o = E/H. \qquad (3.41a)$$

$$j = 2.137\ 384\ 084\ \underline{83} \times 10^{-30}\ Wb\ m;\ \{\#\}. \qquad (3.41b)$$

Natural unit of electric vector potential:

$$b = e/S = V\varepsilon_o = 4.524\ 480\ 319\ \underline{39} \times 10^{-6}\ C\ m^{-1};\ \{\#\}. \qquad (3.42)$$

Natural unit of magnetic vector potential:

$$y = \phi/S = i\mu_o = 1.704\ 508\ 888\ 98 \times 10^{-3}\ Wb\ m^{-1};\ \{\#\}. \qquad (3.43)$$

Natural unit of electric conductivity:

$$\sigma = G_r/S = H/V = 7.495\ 969\ 013\ \underline{55} \times 10^{10}\ S\ m^{-1};\ \{\#\}. \qquad (3.44)$$

Natural unit of electric resistivity:

$$\rho = 1/\sigma = Z_o S = V/H = 1.334\ 050\ 338\ \underline{51} \times 10^{-11}\ \Omega\ m;\ \{\#\}. \qquad (3.45)$$

Natural unit of voltage (electric potential) density:

$$Y = V/A = 4.075\ 085\ 005\ \underline{53} \times 10^{32}\ V\ m^{-2};\ \{\#\}. \qquad (3.46)$$

Natural unit of current (magnetic potential) density:

$$J = i/A = \sigma E_s = 1.081\ 698\ 196\ \underline{27} \times 10^{30}\ A\ m^{-2};\ \{\#\}. \qquad (3.47)$$

Natural unit of compressibility:

$$\beta = 1/P_r = A/\mathrm{F} = V_0/E \qquad (3.48\mathrm{a})$$

$$\beta = 5.423\ 687\ 069 \times 10^{-28}\ \mathrm{Pa}^{-1};\ \{\#\}. \qquad (3.48\mathrm{b})$$

Natural unit of particle fluence:

$$\Phi = 1/A = J/i = B/\phi \qquad (3.49\mathrm{a})$$

$$\Phi = 7.974\ 743\ 056\ \underline{57} \times 10^{26}\ \mathrm{m}^{-2};\ \{\#\}. \qquad (3.49\mathrm{b})$$

Natural unit of moment of inertia:

$$I = m_e S^2 = 1.142\ 279\ 077\ \underline{75} \times 10^{-57}\ \mathrm{kg\ m}^2;\ \{\#\}. \qquad (3.50)$$

Natural unit of angular acceleration (gravitational potential density):

$$\alpha_0 = 1/LC = \omega/t = V_g/A;\ \{\$\}. \qquad (3.51\mathrm{a})$$

$$\alpha_o = 7.167\ 341\ 621\ \underline{19} \times 10^{43}\ \mathrm{r\ s}^{-2};\ \{\#\}. \qquad (3.51\mathrm{b})$$

Natural unit of angle of rotation:

$$\theta = \omega/f = \lambda_c/S = 2\pi_{\mathrm{qed}} = 1/2\alpha;\ \{\$\}. \qquad (3.52\mathrm{a})$$

$$\theta = 68.517\ 999\ 8395\ \mathrm{r\ c}^{-1};\ \{\#\}. \qquad (3.52\mathrm{b})$$

Where $\alpha = 7.297\ 352\ 5376\ (50) \times 10^{-3}$ is the fine-structure constant.

Natural unit of (linear) time:

$$\tau = T/\lambda_c = \mathrm{D/H} = B/E_s;\ \{\$\}. \qquad (3.53\mathrm{a})$$

$$\tau = 3.335\ 640\ 951\ 981\ \underline{52} \cdots \times 10^{-9}\ \mathrm{s\ m}^{-1};\ \{\$,\ \#\}. \qquad (3.53\mathrm{b})$$

Natural unit of circular wave number:

$$\Lambda = 1/S = E_s/V = H/i = g/V_g = 2\pi_{qed}/\lambda_c; \{\$\}. \qquad (3.54a)$$

$$\Lambda = 2.823\ 958\ 756\ \underline{17} \times 10^{13}\ r\ m^{-1}; \{\#\}. \qquad (3.54b)$$

Natural unit of intensity:

$$I_o = P/A = E_s H = 5.527\ 465\ 987\ \underline{14} \times 10^{35}\ W\ m^{-2}; \{\#\}. \qquad (3.55)$$

Natural unit of angular momentum: L_p. Also see the natural unit of action, equation (3.22a).

$$L_p = I\omega = \phi e = h/\theta. \qquad (3.56a)$$

$$L_p = 9.670\ 552\ 226\ 85 \times 10^{-36}\ J\ s; \{\#\}. \qquad (3.56b)$$

Rydberg constant:

$$R_\infty = 1/2\lambda_c(4\pi_{qed})^2 = 1/S(4\pi_{qed})^3; \{\$\}. \qquad (3.57a)$$

$$R_\infty = 10\ 973\ 731.568\ 527\ m^{-1}. \qquad (3.57b)$$

Natural unit of flux of the gravitational field:

$$\Phi_G = gA = m_e/z_o; \{\$\}. \qquad (3.58a)$$

$$\Phi_G = 3182.607\ 312\ \underline{42}\ J\ kg^{-1}\text{-}m; \{\#\}. \qquad (3.58b)$$

In equation (2.3a), we have calculated the natural unit of energy. Here in equations (3.59)-(3.61), we corroborate the natural unit of energy by means of additional relationships:

$$E = m_e V_g = P_r V_o = fh = Hj = Pt. \qquad (3.59)$$

$$E = Ve = \eta Q = \phi i = FS = \hbar_e/t. \qquad (3.60)$$

$$E = m_e c^2 = e^2/C = \phi^2/L = V_o/\beta = 8.187\ 104\ 715 \times 10^{-14}\ J. \qquad (3.61)$$

There is another very important point here, which concerns the c^2 in Einstein's (1905) formula from special relativity, $E = m_e c^2$, equation (3.61), which in reality is V_g, equation (3.36a). Therefore $E = m_e V_g$, equation (3.59), enables us to connect three independent laws: the conservation of energy, the conservation of mass, and gravitation. In addition, in view of the urgent need to find new, renewable, and environmentally friendly alternative energy sources, $E = m_e V_g$ can facilitate generation of energy via the field of gravitation. [10]

References

[1] Bedrij O (1993) Scale Invariance, Unifying Principle, Order and Sequence of Physical Quantities and Fundamental Constants. *Proc Inst Math Natl Acad Sci Ukraine*; also, (1994) *Ukrainian Mathematical Journal,* tr of the Proc Inst Math Natl Acad Sci Ukraine (Allerton Press, New York, NY), pp 67-73.

[2] Weinberg S (1992) *Dreams of a Final Theory* (Pantheon Books, New York), p 204.

[3] Peebles P J E (1993) *Principles of Physical Cosmology* (Princeton University Press, Princeton, NJ), pp. 134-154.

[4] Mohr P J, Taylor B N (2005) CODATA Recommended Values of the Fundamental Physical Constants: 2002. *Rev Mod Phys* vol. 77(1), pp. 1-107.

[5] Dirac P (1928) The Quantum Theory of the Electron. *Proc R Soc Lond* **117A**, 610-624; **118A**, 351-361.

[6] Dirac P (1967) *The Principles of Quantum Mechanics* (Clarendon Press, Oxford).

[7] Fushchych W, Zhdanow R (1997) *Symmetries and Exact Solutions of Nonlinear Dirac Equations.* (Mathematical Ukraina Publisher).

[8] Bedrij O, Fushchych W I (1993) Fundamental Constants of Nucleon-Meson Dynamics. *Proc Inst Math Nat Acad of Sci Ukraine* N 5.

[9] Bedrij O, Fushchych W I (1995) Planck's Constant is not Constant in Different Quantum Phenomena. *Proc Inst Math Nat Acad Sci Ukraine* N 12.

[10] Bedrij O (2002) New Relationships and Measurements for Gravity Physics. Fourth Inter Conf: Symmetry in Nonlinear Math Phys. *Proc Inst Math Natl Acad Sci Ukraine* vol. 43, Part 2, pp. 589-601.

[11] Lorentz H A (1904) Electromagnetic Phenomena in a System Moving with Any Velocity Less Than That of Light Proc Royal Acad Amsterdam 6:809-830. (Reprinted 1952 in *The Principle of Relativity,* tr Perrett W, Jeffery G B, pp 9-34. Dover: New York).

[12] Bedrij O, Fushchych W I (1991) On the Electromagnetic Structure of Elementary Particles Masses. *Proc Inst Math Nat Acad Sci Ukraine* N 2, pp 38-40 (in Russian).

[13] Bedrij O (1994) Connection of π with the Fine Structure Constant *Proc Inst Math Natl Acad Sci Ukraine* N 10

4

The Mathematization of Physics

Every darkness can be lighted.

—John Archibald Wheeler, 1911-2008

At its most fundamental level—the empty space—we live, move, have our being, become conscious of ourselves, and rise to a new birth of freedom.

—Orest Bedrij, 1933-20xx

Abstract

The computationally classifiable methods for rigorous mathematical derivation of experimentally observable results in physics, as provided in papers 1-3, have been given. Now we describe how to produce the natural spectrum of physical quantities, the qualitative content of equations, the orders of infinity, and how to make a new fundamental tool for the experimentally observable mathematization of physics via the LSPR.

The LSPR is a powerful research tool used for the prediction, verification, and the unification of the laws of physics, symmetries, and the fundamental physical constants. It will enable scientists and engineers to determine as yet unseen dimensions of new physical laws, and their fundamental physical constants, and to facilitate many paradigm-shifting discoveries. Also, it will help to predict the results of measurement and to show interdisciplinary and multidisciplinary connections of fields, empowering physics by extending our knowledge of the laws of nature, thus promoting excellence in science, engineering, education, and technology.

1 Introduction

In this paper we will discuss how to construct the logarithmic slide rule of physical relationships (LSPR). The LSPR is a simple and experimentally observable mathematical tool for understanding computable elements and the dimensional uniformity of physical equations, the order of the quantities, and the prediction of the laws of physics, constants, and symmetries (table 4.1). [1, 2]

Furthermore, it will help in examining the deepest levels of computationally classifiable physics [3, 4] and organizing principles [5, 6] in a rational, systematic, and rigorously precise formalism. Also, it can assist in understanding the hierarchy problem in the causal nexus of physics and piece together computability of constants (i.e., equations (2.1c) and (2.1d)). It can also simplify constants (i.e., equations (1.9a)-(1.13a)) and discover new knowledge (i.e., equations (2.1)-(3.61)).

The LSPR is applicable for scientist-to-scientist communication. It can foster international collaboration in this field and has a potential to make physics more accessible in education and to the nonspecialist. In addition, the LSPR can be applied to improve the quality, consistency, management, accessibility, and education of fundamental issues and to advance science and technology for the betterment of planetary life.

2 The Logarithmic Slide Rule of Physical Relationships

One can equate the LSPR, [7] partially reproduced on '1' book cover and as a Spectrum of Physical Quantities in Appendix I, to a logarithmic slide rule used in the mathematics of science and engineering. While a logarithmic slide rule enables one to rapidly calculate (multiply, divide, and predict) *relationships of numbers*, the LSPR enables one to rapidly calculate (multiply, divide, and predict) *relationships of physical quantities*, the fundamental constants, and the laws of physics.

In an ordinary slide rule one can extract square roots, calculate trigonometric functions and logarithms. With the LSPR, one can do similar things with the laws of physics, physical quantities, and the fundamental constants. The slide rule enables one to produce solutions by continuous operations, without the need for intermediate readings, thus giving added speed and flexibility to the solution of many problems. One can do the same with the LSPR.

Instead of a natural number sequence on a logarithmic slide rule, we have a sequence of natural quantities on the LSPR. On a logarithmic slide rule, we position the ordering of numbers logarithmically in reference to a *mathematical zero*. On the LSPR however, we position the ordering of physical quantities logarithmically in reference to the rest frame—the *natural zero* (the zero velocity frame of reference, logarithmic zero: $1 = 10^0 = e^0 = x^0$) as found in the exponent. [7]

On the slide rule and the LSPR, adding logarithms serves to multiply, and subtracting logarithms serves to divide. Since we understand the values and the *ordering of numbers*, it is not problematic to relate a numerical communication or to balance your checkbook. Given that most of us do not understand the "qualitative meaning of quantities," we find it challenging to predict the ordering of the quantities or their relationships.

As Dmitriy Mendeleyev (1869), John Newlands (1864), Lothar Meyer (1864), and others classified the elements according to their atomic mass, the LSPR classifies ordering of physical quantities according to their fundamental physical constants. While the periodic table is a tremendously profound scientific tool, it does not predict the synthesis of the elements.

Equations (1.9a)-(3.61) are a condensed list containing the values of constants and most frequently used conversion factors. To construct the LSPR, [8, 9] it is necessary to incorporate constants missing from the body of physical laws to our abbreviated listing, shown in papers 1-3. Next, to make the listing more comprehensive, applying reciprocal relations in the one-and-the-many principle, paper 1, we need to *invert* each fundamental physical constant. This will yield a mirror image of each quantity, which is the *same* natural logarithm that employs the transcendental nature base e where $e = 2.71828\ldots$, but with an opposite sign. By way of illustration, the natural unit of conductance ($G_r = 2.6544 \times 10^{-3}$ S), equation (2.9b), and the natural unit of resistance ($Z_0 = 376.73\ \Omega$), equation (2.11b), would be -5.9315 for $q_k\ G_r$ and 5.9315 for $Q_k\ Z_0$.

Next, we place each constant on a paper in reference to the '1' by means of the natural log scale. The natural log scale will extend from $\ln -131.114\ 3248$ ($I = 1.142\ 279\ 153 \times 10^{-57}$) as shown in equation (3.50) to $\ln 100.980\ 6938$ ($\alpha_0 = 7.167\ 341\ 443 \times 10^{43}$) as shown in equation (3.51b) or a spectrum extending to 10^{100} power. The natural log scale will generate the *natural spectrum of physical quantities*, upon which we can easily expand as new quantities and constants are discovered or predicted. On the left-hand side of '1', we will have q_k (–∞) quantities, and on the right-hand side of '1', we will have Q_k (+∞) quantities. By combining *three* natural spectra of quantities (scales A, B, and C), we can construct the LSPR. Scale A is *inverse* to scales B and C and is used in multiplication, while scales B and C are used in division of quantities.

Transformations from one quantity (or infinity) to another are achieved by shifting (translating, exponentiating) the frame of reference, of 10^0 power, to another frame of reference, having a new equilibrium center of 10^{-n} or 10^n power. Note that all laws of physics are in equation equilibrium. That is, the left side of the equation equals the right side of the equation.

Just as the equal sign in physical relationships allows one to go from one system of measurement to another, so the '1' on the LSPR allows one to shift the equilibrium center from one frame of reference (quantity and symmetry) to another. **The equilibrium center defines the frame of reference.** Because not all-physical relationships between quantities are presently known, with the LSPR we can obtain new (previously unknown) relationships, as well as find new fundamental physical constants.

To multiply (or divide) quantities, we align the multiplicand (or numerator) quantity on scale A (or scale B for division) with the multiplier (or denominator) quantity on scale C and read the answer on scale A (or scale B for division) over '1' on scale C.

To predict physical relationships for a quantity, or to solve for multiple quantity relationships, i.e., equations (3.59)-(3.61), we align '1' of scale C with the desired quantity on scale A for multiplication (or scale B for division) and note the quantities that *line up* with themselves. These resulting quantities are the product (or quotient) of the desired quantity.

To verify the validity of our relationships, explore new connections in the deep entanglement of fields, or to test new laws of physics, we multiply (or divide) our resultant quantity by the constants listed in equations (1.9b)-(3.58b). Both sides of the equation should be equal.

3   The Orders of Infinity

The depth and richness of infinity lie deep within the foundations of mathematics and physics. In the search for understanding, similar to the challenge of grasping the concept of zero, mathematics and physics has had the challenge of grasping the deep nature of infinity, more specifically Absolute and *actual* infinity. Infinities, which come in different sizes, are an essential part of physics, yet infinities have been banished or avoided from general relativity, quantum theory, and the esoteric theory of strings.

A small number of individuals have been given a preview of infinity. Leonhard Euler, Carl F. Gauss, Gottfried Leibniz, Isaac Newton, and others in their research within the foundations of mathematics *approached* infinity. They were dealing with the *potential* infinity of limits. In quantum field theory, Werner Heisenberg, Wolfgang Pauli, Julius R. Oppenheimer, and Ivar Waller also encountered the problem of potential infinity in physics.

Galileo Galilei, and more so Georg F. L. P. Cantor, went the next step up—they ascended from potential infinity to properties of *actual* infinity. Cantor, whom Bertrand Russell described as one of the greatest intellects of the nineteenth century, succeeded in constructing one of the most remarkable theories in the annals of humanity—the idea of actual infinity. Cantor called his infinities *transfinite numbers*, except the ultimate level of infinity that he called the Absolute, our '**1**'. While working with transfinite numbers, Cantor discovered transfinite arithmetic—the mathematical laws of transfinite numbers.

The concept of dimensions is central to physics and mathematics. Cantor wanted to unlock the mysterium magnum of nature via the orders of infinity of various objects of *different dimensions* with traditional mathematical formalism. As his work led to the foundations of mathematics, Cantor discovered that dimensions were irrelevant at the Absolute. That is true. We found that every physical quantity has its own dimension that when cross multiplied in the physical relationship of '**1**' becomes dimensionless [7] or Appendix I. Indeed, in '**1**' every scale invariant relationship can be considered as an algebraic relation for dimensionless quantities, actual infinities, with '**1**'→0, the simple point, as the central pivot and the least common denominator of all physical relationships in '**1**'→Absolutely Infinite, the empty space.

Also, at '**1**', as we have seen earlier, all dimensions and the laws of physics break down. However, at '**1**' there is the world of quantum potentialities and an order of the fundamental physical constants. Their dimensionless *relationship*, like the laws of physics, can be determined by measurements that Cantor and other mathematicians working on this problem had overlooked.

Cantor also realized that at '**1**' all continuous spaces, lines, planes, or higher surfaces had an identical *order of infinity*. Yet through the operation of *exponentiation* one can change the value of the infinities. Similarly in physics, through the operation of transformation one can change the nature of the physical quantity.

As dimensions were of no significance for Cantor, he tried to solve an impossible mathematical problem—the *order* of his transfinite cardinal numbers or the orders of infinity by way of mathematics. He wanted to tag them sequentially, without the fundamental constant measurement of physics, as we did to produce the spectrum of physical quantities in the LSPR.

The challenge concerning the different orders of infinity is known as the continuum hypothesis (which was generalized by Felix Hausdorff) and its extension.

Cantor was certain that the continuum hypothesis was true. And it is! However, without the measurement of the fundamental physical constants, Cantor's orders of infinity (the well-ordering principle) could not be proved. The continuum hypothesis is undecidable in mathematics, i.e., there is no solution within our system of mathematics to his problem. Here, mathematics needed physics to solve the long-standing ordering of infinities or distinct objects. In essence, one-winged birds do not fly; flight is only possible with two wings.

While Cantor's monumental work on infinity focused on the nature, the sizes, and the infinite hierarchical levels of sets, our work on infinity focuses on the nature, the sizes, and the infinite hierarchical levels of the physical quantities. Whereas dimensions were immaterial for Cantor, we discovered that the fundamental physical constants, which are dimensionless, are indispensable in the determination of the landscape of '1'. An equation without the '1' has no meaning. The reason something is unchanging it is because it is infinite. We can repeat an experiment an infinite number of times and an equation should remain unchanging. With measurement, constants, exponentiation, or transformation from one quantity to another in '1', we could "distinguish quantities." [7]

Cantor concluded that there was an Absolute that could not be understood or tested within the foundations of mathematics. However, with the measurements of physics and astrophysics today, we can prove that '1' exists and extends beyond the confines of a physical quantity, set, or galaxy.

Given that each point can have an infinite number of points within itself, *the value of each physical quantity in '1' can extend from zero to infinity*. Thus, we can regard every q_k ($-\infty$) and Q_k ($+\infty$) quantity as relatively or actually infinite in '1'.

Additionally, as every physical quantity is an expression and "renormalizable" in '1' (canceling the infinities in a renormalization), we can consider the natural spectrum of physical quantities as the *natural spectrum of infinities*. The continuum hypothesis—which Cantor, J. von Neumann, K. Gödel, D. Hilbert, E. Zermelo, A. A. Fraenkel, S. Banach, P. Cohen, and other great minds worked so hard to answer—can now be solved via the mathematization of physics and the LSPR.

Visualization can assist students learning some of the most demanding principles in the physics curriculum. One of the key elements of the LSPR is that it visually demonstrates, with lucid exactness and with a unifying precision, how physical quantities and the laws of physics are interrelated and classified, and how the same essence—the '**1**'—appears in different forms.

Further, the different forms—physical quantities and constants—once decoded and organized, like the table of elements, have an orderly and logical structure. For instance, it is well-known that when an area bounded by a curve is *rotated* about an axis, a corresponding volume is produced. In a similar way, other physical quantities or quantum states can be generated and predicted with the LSPR. Listed below are a few examples from the left-hand side (q_k quantities) of the LSPR:

$$\text{Volume: } V_o \text{ equation (3.6b)} = \text{Angular area:} \tag{4.1}$$
$$A \text{ equations (3.5b) x } S \text{ (3.4b).}$$

$$\text{Capacitance: } C \text{ equation (3.7b)} = \text{Angular permittivity:} \tag{4.2}$$
$$\varepsilon_o \text{ equations (1.10b) x } S \text{ (3.4b).}$$

$$\text{Self-inductance: } L \text{ equation (3.8b)} = \text{Angular permeability:} \tag{4.3}$$
$$\mu o \text{ equations (1.9b) x } S \text{ (3.4b).}$$

$$\text{Gravitance: } \Omega \text{ equation (3.39b)} = \text{Angular penetrability:} \tag{4.4}$$
$$z_o \text{ equations (3.38b) x } S \text{ (3.4b).}$$

$$\text{Mag. flux quantum: } \Phi_o \text{ equation (3.17)} = \text{Flux mag. induc. } \phi \tag{4.5}$$
$$\text{equations (3.15) x } \pi_{qed} \text{ (1.11b).}$$

All quantities on the left-hand side, q_k, of '**1**', in the LSPR, are the inverse of the right-hand side, Q_k, of '**1**'. Hence, to produce an angular quantity, when going from q_k to Q_k quantities, instead of multiplication by S, we would divide the quantity by S. Listed below are a few examples from the Q_k part of the LSPR:

$$\text{Voltage density: Y equation (3.46)} = \text{Angular electric field str.:} \tag{4.6}$$
$$E_s/S \text{ equation (3.11).}$$

$$\text{Current density: J equation (3.47)} = \text{Angular magnetic field str.:} \tag{4.7}$$
$$H/S \text{ equation (3.12).}$$

Electric field str.: E_s equation (3.11) = Angular electric potential:

V/S equation (2.13b). (4.8)

Magnetic field str.: H equation (3.12) = Angular magnetic potential: i/S equation (3.1). (4.9)

4 Symmetries in the Electric, Magnetic, and Gravitation Relationships

In table 4.1 we call your attention to a sample of symmetries between the electric, magnetic, and gravitation relationships and constants of equations (1.9a)-(3.58a).

TABLE 4.1. Symmetries in the Electric, Magnetic, and Gravitation Relationships.

	Electric	Magnetic	Gravitation
1.	$E_s = F/e$ (3.20a)	$H = F/\phi$ (3.20a)	$g = F/m_e$ (3.35a)
2.	$D = e/A$ (3.13)	$B = \phi/A$	$M = m_e/A$ (3.33a)
3.	$V = E_s S$ (3.11)	$i = HS$ (3.40)	$V_g = gS$ (3.36a)
4.	$C = e/V$ (3.7a)	$L = \phi/i$ (3.16)	$\Omega = m_e/V_g$ (3.39a)
5.	$\varepsilon_0 = D/E_s$ (3.13)	$\mu_0 = B/H$ (3.14)	$z_0 = M/g$ (3.39a)
6.	$1/\varepsilon_0 = S/C$ (3.7a)	$1/\mu_0 = S/L$ (3.8a)	$G_e = S/\Omega$ (3.37a)
7.	$E = eV$ (3.6)	$E = \phi i$ (3.60)	$E = m_e V_g$ (3.59)
8.	$F = eE_s$ (3.20a)	$F = \phi H$ (3.20a)	$F = m_e g$ (3.35a)
9.	$S = C/\varepsilon_0$ (3.7a)	$S = L/\mu_0$ (3.8a)	$S = \Omega/z_0$ (3.39a)
10.	$\Phi_E = e/\varepsilon_0$ (3.18)	$\Phi_H = \phi/\mu_0$ (3.18)	$\Phi_G = m_e/z_0$ (3.58a)
11.	$Y = V/A$ (3.46)	$J = i/A$ (3.47)	$\alpha_0 = V_g/A$ (3.51a)
12.	$\rho_c = e/V_0$ (3.25)	$? = \phi/V_0$	$d = m_e/V_0$ (3.26)
13.	$A = \Phi_E/E_s$ (3.18)	$A = \Phi_H/H$ (3.19)	$A = \Phi_G/g$ (3.58a)
14.	$P_r = DE_s$ (3.21a)	$P_r = BH$ (3.21a)	$P_r = Mg$ (3.35a)
15.	$e = CV$ (3.15)	$\phi = Li$ (3.16)	$m_e = \Omega V_g$ (3.39a)
16.	$b = V\varepsilon_0$ (3.42) y	$y = i\mu_0$ (3.43)	$\mu = V_g z_0$ (3.34a)
17.	$1/E_s = e/F$ (3.20a)	$1/H = \phi/F$ (3.20a)	$1/g = m_e/F$ (3.35a)

If we use the five high-precision constants, equations (1.9b)-(1.13b), and their derivative constants, equations (2.1b)-(3.58b), in table 4.1 relationships, they all cross-check and are in agreement with each other to nine decimal places. These were determined by means of the LSPR. Predicted new gravitation relationships are highlighted with bold letters; equation references and constants are shown in brackets.

Notice many commonly used physical quantities do not exist in inverted form. With equations (1.9b)-(3.58b) constants and instrumentation available, it should not be a difficult process to characterize and measure them. As cases in point, the electric field strength E_s, equation (3.11), the magnetic field strength H, equation (3.12), and the acceleration of gravity g, equation (3.35a), are Q_k quantities. Through inversion, we can characterize their constants and verify their q_k counterparts with measurements.

Nature loves to hide; she does not easily surrender her solutions to complex challenges. With the LSPR, like a calculator in mathematics, the science of physics is revealed as being logical, elegant, and of crystalline clarity at the fundamental level, allowing for the development of new disciplines and resulting in the consequent emergence of entirely new fields of science. In addition, with a mechanical version or a software form of the LSPR, we can have in our hands an extremely sophisticated tool for predicting the results of high-precision measurements. This can help to creatively study physical concepts and principles at a deeper level, to reason analytically in a careful logical progression, as well as to accelerate the advancement and diffusion of the knowledge of physics, technology, and education, and their application to human welfare.

5 Concluding Remarks

Although emotional as well as intellectual considerations are significant impediments to progress, a number of conclusions relevant to this work may be drawn.

1. The idea of reproducible experiments, a cornerstone of science, and the discovery of the fundamental physical constants and the laws of physics are incontestably the most noteworthy attainments of scientific thought that have survived an impressive variety of nontrivial tests. Like minuscule seeds that conceal from view sequoia trees and hidden dimensions, so does each background independent algorithmically compressed dimensionless point conceals dimensions from which

all physics, awareness, and universes emerge. We characterize the "dimensionless point," equivalently, the "zero point," as follows:

$$(q_k, \text{dimensionless point, '}1\text{'}) =$$
$$1/(Q_k, \text{dimensionless empty space, '}1\text{'}). \qquad (4.10)$$

2. Using '1' as our background independent "rest frame" of nature, we have formulated a relatively straightforward, experimentally supported approach to advance the mathematization of physics and, as John Wheeler put it, to "derive everything from nothing, all law from no law." [10] This concept has a fundamental application in physics, cosmology, engineering, technology, and mathematics. The current experimental evidence shows that this approach provides an observationally successful account of the otherwise incomplete relativistic physics and cosmology. The results of every experiment (the workhorses of science), at this point in time, agree with the five high-precision constants, equations (1.9b)-(1.13b), and the CODATA internationally recommended values of the fundamental constants, equations (2.1b)-(2.17b).

There is a strong empirical case that the physical relationships in equations (3.1)-(3.61) are consistent with evidence. The hidden dimensions of equations (3.1)-(3.61) in effect, through the '1' and the laws of physics, are connected to observation. Thus, in view of the impressive body of evidence, both experimental and theoretical, that supports the validity of the relationships, they have the potential to lay the groundwork for advancing physics, and the precision of the CODATA constants, without resorting to massive measurements.

3. As a result of the '1', the scientific effort is a collective and self-correcting process. Although the book of nature has many unexpected alternatives up her sleeve, in reference [11] we see the exceeding beauty, simplicity, and mathematical depth in the predictive power of the '1'. Once the QED constants are determined and used, to predict a similar spectrum of constants in nucleon-meson dynamics, we no longer require five constants but *one* (i.e., proton or neutron mass).

4. The rapid advancement of knowledge, even more so than the compounding of interest of capital, increases at an accelerating rate. Until now physics has remained mostly an experimental science. We have shown that the foundation of physical science not only is

unchanging and the starting point from which all explanations may be traced, but also via the mathematization of physics can be used to predict the manner in which the laws of physics have a code-script, order, structure, organization, and scale. Using Erwin Schrödinger's simile, "they are architect's plan and builder's craft—in one." This means that the scientific endeavor of physics (new theories, new discoveries, and new understanding) can accelerate to greater speed, producing a more rapid rate of advancement without huge outlays of capital.

5. Where does the evidence seem to be leading? Although this is not the focal point of this work, we have observed that in the standard picture, the foundation of the CBR is not restricted to a small area, but is evenly and isotropically dispersed throughout the empty space. That is, the radiation is strikingly smooth and uniformly razor-sharp in all directions.

Given that nature is quite capable of surprising us, the world of quantum potentialities can be at any point in the universe. In addition, since pure awareness is also '1', it would suggest that von Neumann's "quantum jumps," called events (creations), can take place all over the empty space, and according to the Schrödinger equation, continually, and therefore do not have to be in one localized explosion as the standard hot evolving cosmological model, the hot big bang, suggests.

6. The works of Albert Einstein, Richard P. Feynman, and Niels Bohr, three of the world's eminent physicists of the last century, successfully combine intellectual perceptiveness and engaging humanity with extraordinary brilliance and creative productivity, forming a powerful marriage. Yet their insights that quantum physics is too paradoxical ever to be reconciled with a conventional grasp (Einstein) or "I can safely say that nobody understands quantum mechanics" (Feynman), or "Anyone who is not shocked by the quantum theory does not understand it" (Bohr), may no longer be valid. This is particularly fitting when we consider the cumulative aspects of science in the experimentally observable mathematization of physics.

To elicit the best thinking, testable predictions, and achieve the experimentally observable mathematization of physics we suggest the following: (1) the natural zero in physics, '1'; (2) scale invariance; (3) the one-and-the-many principle; (4) the natural spectrum of physical quantities and the laws of physics; (5) the LSPR; and (6) the π_{qed} framework of physics, equations (1.9a)-(1.13a), (2.1a)-(2.18a), and (3.52a).

Thank you.

References

[1]   Fushchych W I, Nikitin A G (1987) *Symmetries of Maxwell's equations* (D Reider Publishing Company, Dordrecht).

[2]   Fushchych W I, Nikitin A G (1994) *Symmetries of Equations of Quantum Mechanics* (Allerton Press, New York, NY).

[3]   Jammer M (1961) *Concepts of Mass in Classical and Modern Physics* (Harvard University Press, Cambridge, MA).

[4]   Weinberg S (1967) Conceptual Foundations of the Unified Theory of Weak and Electromagnetic Interactions. *Rev Mod Phys* 52:255-77.

[5]   Feynman R P (1965) *The Character of Physical Law* (MIT Press, Cambridge, Mass).

[6]   Greenberger D M, Zeilinger A eds (1995) *Fundamental Problems in Quantum Theory: A Conference Held in Honor of Professor John A. Wheeler* (N Y Acad Sci New York).

[7]   Bedrij O (1993) Scale Invariance, Unifying Principle, Order and Sequence of Physical Quantities and Fundamental Constants. *Proc Inst Math Natl Acad Sci Ukraine*; also, (1994) *Ukrainian Mathematical Journal,* tr of the Proc Inst Math Natl Acad Sci Ukraine (Allerton Press, New York), pp 67-73.

[8]   Trylis O (Dec 2001, Feb 2002) Cosmolog: Slide Rule That Knows Physics. *Physics: Schkilnyj Swit, Kyiv*, No. 36 (120) p. 9; No. 4 (124) pp. 3-4, 9.

[9]   Bedrij O (1993) *Cosmolog Logarithmic Slide-Rule of Physical Relationships*, U.S. Copyright Office, TX 3 705 977.

[10]   Wheeler J A (1994) *At Home in the Universe* (AIP Press, Woodbury, NY), p 302.

[11]   Bedrij O, Fushchych W I (1993) Fundamental Constants of Nucleon-MesonDynamics *Proc Inst Math Nat Acad of Sci Ukraine* N 5.

APPENDIX I

UDC 517.9:519.46

O. BEDRIJ

SCALE INVARIANCE, UNIFYING PRINCIPLE, ORDER AND SEQUENCE OF PHYSICAL QUANTITIES, AND FUNDAMENTAL CONSTANTS

(Submitted by Corresponding Member W. I. Fushchich)

Scale invariance is a fundamental concept in physics. The concept of scale invariance can be expanded to enable us to more fully articulate fundamental questions concerning the scale, sequence, and classification between different dimensional quantities. Scale invariance permits us to shed new light on the innermost structure, the deep and profound order, and the mutual interaction of the fundamental constituents of physics. Also, scale invariance helps us derive new laws of nature and fundamental physical constants.

When relationships between different physical quantities are expressed by an equation with an equal sign (=), it means that the corresponding physical quantities are in equilibrium. For example,

$$q_1 = q_2 q_3, \text{ or } 1 = q_1/q_2 q_3 = q_2 q_3/q_1, \ q_1 \neq 0, \ q_2 q_3 \neq 0, \qquad (1)$$

where $q_1 = V$ (electric potential in volts), $q_2 = R$ (resistance in ohms), $q_3 = i$ (magnetic potential in amperes). Formula (1) is the Ohm's law, which shows that q_1 is in equilibrium with $q_2 q_3$. Formula (1) is a simple expression of a more general form:

$$1 = (q_1^{x_1} \cdot q_2^{x_2} \cdot q_3^{x_3} \ldots q_s^{x_s}) / (p_1^{j_1} \cdot p_2^{j_2} \cdot p_3^{j_3} \ldots p_z^{j_z}) \qquad (2)$$

or

$$1 = Y'/KX, \qquad (3)$$

where

$$Y' \equiv (q_1^{x_1} \cdot q_2^{x_2} \cdot q_3^{x_3} \ldots q_s^{x_s}), \qquad (4)$$

$$1/KX \equiv (p_1^{j_1} \cdot p_2^{j_2} \cdot p_3^{j_3} \ldots p_z^{j_z}), \qquad (5)$$

$$(q_s)^0 = 1, \ (q_s^{-1})^0 = 1,$$

$q_1, q_2, q_3, \ldots q_s, p_1, p_2, p_3, \ldots, p_z$ are quantities, $x_1, x_2, x_3, \ldots, x_s, j_1, j_2, j_3, \ldots, j_z$ are real numbers, K is the slope for line $Y' = KX$, $j, s, x, z = 1, 2, 3 \ldots$ We require that formula (2) is scale invariant. That is, formula (2) is invariant with respect to the following transformations:

$$q_1 \rightarrow q_1' = aq_1, \ q_2 \rightarrow q_2' = aq_2, \ q_3 \rightarrow q_3' = aq_3, \ldots, \qquad (6)$$

$$p_1 \rightarrow p_1' = ap_1, \ p_2 \rightarrow p_2' = ap_2, \ p_3 \rightarrow p_3' = ap_3, \ldots \qquad (7)$$

Where «a» is scale transformation parameter, and all physical quantities (q_s and p_z) have to be subject to transformation. Hence, based on formula (2), it follows that «1» is always invariant with respect to scale transformation (6) and (7).

When a structure of physical relationships requires a more complete description, we can expand formula (2) to the following formulae:

$$1 = Y'/(KX + C), \qquad (8)$$

$$1 = Y'/(B'X^2 + KX + C), \qquad (9)$$

$$1 = Y'/(D'X^3 + B'X^2 + KX + C), \qquad (10)$$

$$1 = Y^2/(B'X^2 + KX + C), \qquad (11)$$

$$1 = (A_2Y^2 + A_1Y')/(B'X^2 + KX + C), \qquad (12)$$

where K, C, B', D', A_1, A_2 are real parameters.

Please note that every physical quantity of equation (2) has its own dimension that, when cross multiplied in (2), becomes dimensionless. Indeed, we have a relationship for each physical quantity as it relates to all other quantities in (2). Further, we can consider all these relationships as algebraic equations to define the relative dimensionless value of each physical quantity.

For example, let us consider equation (2) and the following relationships:

$$1 = q_1/q_2 \cdot q_3 = q_2 \cdot q_4 = q_5 \cdot q_6 = q_7 \cdot q_8, \qquad (13)$$

where $q_4 = G$ (conductance in siemens), $q_5 = T$ (period of harmonic motion in seconds/cycle), $q_6 = f$ (frequency of the motion in cycles/second), $q_7 = \lambda$ (wavelength in meters/cycle), $q_8 = n$ (wave number in waves or cycles/meter).

From formula (2) or (13) we can see that in cross multiplication, dimensions disappear. Then relationships (2) or (13) can be considered as an algebraic relation for dimensionless quantities, with «1» as the central pivot and the least common denominator of all physical relationships (2)-(13). Further, «1» can also be viewed as a product of quantity and its reciprocal ($q_5q_6 = q_7q_8 = 1$).

From formulae (1), (2), and (8)-(13), we see that «1» can be viewed as the ultimate unifying principle (Absolute) between all physical quantities and relationships of type (2). Further, that «1» is the scale invariant equilibrium frame of reference between every physical relationship. In the logarithmic or natural log scale (see Table 1), the equilibrium frame of reference «1»→0 ($1 = 10^0 = e^0$). Therefore, as in numbers, so in physics, we can consider zero the starting frame of reference for the scale of all physical relationships.

The concept of «0» (zero) is well-known and understood in mathematics. However, in physics, the concept of zero has not been adequately defined mathematically or discussed as it relates to the scale of invariance for the laws (quantity relationships) of physics and the fundamental physical constants. Even in the vacuum idea in quantum field theory, «zero» is not a complete zero concept because the quantum vacuum exists in time and space.

In numbers, while direct understanding of the zero concept came into being only recently (approximately in the ninth century), its implied value was always a part of every rational number. That is, the frame of reference of every number on a mathematical scale is «0» (zero).

We propose that just as each number, on a mathematical scale, has a unifying principle (zero) as its starting frame of reference, so each physical quantity, on the physics scale of quantities, also has a unifying principle (equilibrium frame of reference—zero) as its starting frame of reference. Further, «1» includes within itself all physical quantities and serves as the ultimate principle out of which all physical quantities and relationships (reality) emerge (2)-(13).

Equation (2) gives us only relationships between the «1» and physical quantities. To locate those quantities on one scale, we have to know

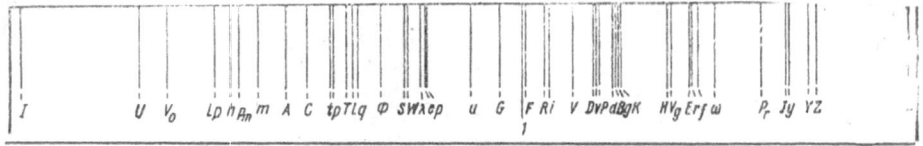

Figure 1: Spectrum of Physical Quantities

their absolute values (constants) with respect to the equilibrium frame of reference. Further, all relationships between the constants (numbers) on the «1» scale must be scale invariant. Hence, the whole spectrum of numbers must have one-to-one correspondence between physical quantities. The challenge is to find such numbers that when cross multiplied by themselves will satisfy formulae (2) and (13). Further, when the laws of physics (physical relationships) are involved, we must find numbers that will satisfy all relationships for all physical quantities (2) that q_1, q_2, . . ., q_8 and p_1, p_2, . . ., p_8 are part of.

At the present, we know of over seventy physical quantities and over three hundred physical relationships that the physics world has directly or indirectly verified. The numbers (constants) that must be found for formula type (2) or (13) must satisfy simultaneously all these relationships in three hundred equations. That is, we must find constants that will be applicable to all quantities of type (2) that have been discovered and verified by physics.

To solve the problem, we have used computers. [1, 2] When we performed the computations, we found that fundamental constants such as Planck's constant, velocity of light, electron charge and mass, Compton wavelength, permittivity and permeability of free space, Rydberg, and the fine-structure constant enter into all physical relations (laws of physics). We also found additional (new) physical relationships and constants. A partial list of the computed constants is shown in Table 1.

When we affix appropriate numbers (constants) with their corresponding physical quantities, from Table 1, on a scale of one axis, we obtain a spectrum of physical quantities—a natural scale of measurement with its fixed order and sequence of physical quantities, with «1» in the center of the spectrum (Figure 1). The spectrum has a colossal range. For example, when we use the electron constant as the frame of reference, the range will extend from 10^{-57} for the moment of inertia to greater than 10^{43} for angular acceleration. [2]

By combining two physical quantity spectra (two Figure 1s), we can construct a logarithmic slide rule of physical relationships (LSPR). Instead of the numbers on a presently used slide rule, we have physical quantities with LSPR. As with numbers that have a sequence on a slide rule, with LSPR we have a sequence of physical quantities. With a mathematical slide rule we can multiply and divide numbers; with an LSPR we can multiply and divide physical quantities.

Transformations from one physical quantity to another are achieved by shifting (translating) the frame of reference, of 10^0 power, to another frame of reference, having a new equilibrium center of 10^n or 10^{-n} power. Note that all fundamental laws of physics are in equational equilibrium. That is, the left side of the equation equals the right side of the equation. Just as the equal sign allows one to go from one system of measurement to another, so «1» on the LSPR allows one to shift the equilibrium center from one frame of reference (quantity) to another. The equilibrium center defines the frame of reference.

Table 1. Fundamental Constants of Quantum Electrodynamics

Symbols	Constants	Natural Log	Relationships of Quantities
I	$1.142280538 \cdot 10^{-57}$	-131.1143236	$I = mS^2$
V_0	$4.440432224 \cdot 10^{-41}$	-92.91523709	$V_0 = m/d$
h	$6.626075438 \cdot 10^{-34}$	-76.39388047	$h = W/f$
m	$9.109389672 \cdot 10^{-31}$	-69.17083217	$m = F/y = V_0 d$
A	$1.253959463 \cdot 10^{-27}$	-61.9434914	$A = \Phi/B = i/J$
C	$3.135382131 \cdot 10^{-25}$	-56.42187627	$C = q/V = q^2/W$

u	$1.700877491 \cdot 10^{-24}$	-54.73089794	$u = iA = W/B$
$t\,(1/\omega)$	$1.181193493 \cdot 10^{-22}$	-50.49034668	$t = q/i = p/F$
$T\,(1/f)$	$8.093300952 \cdot 10^{-21}$	-46.26325028	$T = 1/f = h/W$
$L\,(1/r)$	$4.449913947 \cdot 10^{-20}$	-44.5588171	$L = \Phi/i = S$
q	$1.60217733 \cdot 10^{-19}$	-43.27775323	$q = CV = F/E$
Q	$3.759275896.10^{-19}$	-42.42489041	$Q = Av - V_0 t$
Φ	$6.035887677 \cdot 10^{-17}$	-37.34622365	$\Phi = F/H = BA$
S	$3.5441129005 \cdot 10^{-14}$	-30.97174576	$S = V/E$
W	$8.187111141 \cdot 10^{-14}$	-30.1336302	$W = Pt = FS$
$\lambda\,(1/n)$	$2.426310586 \cdot 10^{-12}$	-26.74464929	$\lambda = v/f$
ε	$8.854187818 \cdot 10^{-12}$	-25.45013057	$\varepsilon = D/E = C/S$
Q_m	$7.712021552 \cdot 10^{-09}$	-18.68048548	$Q_m = Qd = m/t$
	$1.256637061 \cdot 10^{-06}$	-13.5870714	$= B/H$
M	$7.264500922 \cdot 10^{-04}$	-7.227340774	$M = m/A$
$G\,(1/R)$	$2.654418727 \cdot 10^{-03}$	-5.931529583	$G = i/V = Cf$
α_f	$7.297353079 \cdot 10^{-03}$	-4.920243588	$\alpha_f = S/2\lambda = 1/2\theta$
«1»	$1.000000000.10^{0}$	0.000000000	«1» $= GR = Tf$
F	$2.312005898.10^{0}$	0.8381155014	$F = qE$
θ	$6.851799475 \cdot 10^{1}$	4.227096408	$\theta = \lambda/S = \omega/f$
$R\,(1/G)$	$3.767303134 \cdot 10^{2}$	5.931529583	$R = V/I$
i	$1.356405483 \cdot 10^{3}$	7.212593453	$i = P/V$
η	$2.177842587 \cdot 10^{5}$	12.29126021	$\eta = W/Q = EH/Y$
V	$5.109990628 \cdot 10^{5}$	13.14412304	$V = W/q$
R_∞	$1.097373153 \cdot 10^{7}$	16.21101493	$R_\infty = \alpha^3_f/S$
D	$1.277694676 \cdot 10^{8}$	18.66573816	$D = q/A$
V	$2.99792458 \cdot 10^{8}$	19.51860099	$v = H/D = E/B$
P	$6.931219308 \cdot 10^{8}$	20.35671649	$P = iV = Fv$
D	$2.051464635 \cdot 10^{10}$	23.74440492	$d = m/V_0 = P_r/V_g$
B	$4.813463159 \cdot 10^{10}$	24.59726775	$B = E/v$
G	$7.495967321 \cdot 10^{10}$	25.04021612	$g = H/V = G/S$
$\Lambda\,(1/S)$	$2.823958118 \cdot 10^{13}$	30.97174576	$\ddot{E} = E/V$
H	$3.830432276 \cdot 10^{16}$	38.18433915	$H = i/S = F/\Phi$

V_g	$8.987551787 \cdot 10^{16}$	39.03720197	$V_g = W/m$
E	$1.443039952 \cdot 10^{19}$	44.11586873	$E = V/S = F/q$
$r\ (1/L)$	$2.247234468 \cdot 10^{19}$	44.5588171	$r = i/\Phi$
$f\ (1/T)$	$1.235589787 \cdot 10^{20}$	46.26325028	$f = W/h = v/\lambda$
$\omega\ (1/t)$	$8.466013454 \cdot 10^{21}$	50.49034668	$\omega = f\theta$
P_r	$1.843764464 \cdot 10^{27}$	62.7816069	$P_r = F/A = BH$
J	$1.081698032 \cdot 10^{30}$	69.15608485	$J = i/A$
Y	$2.538046983 \cdot 10^{30}$	70.00894767	$Y = F/M = v\omega$
Y	$4.075084387 \cdot 10^{32}$	75.08761443	$y = V/A$
Z	$3.493766483 \cdot 10^{33}$	77.23628844	$Z = \alpha/d$
S_p	$5.527466807 \cdot 10^{35}$	82.30020788	$S_p = P/A = EH$
α	$7.167338381 \cdot 10^{43}$	100.9806934	$\alpha = W/I = 1/LC$

Because not all physical relationships between quantities are presently known, with LSPR we can obtain new (unknown) relationships between physical quantities, as well as find new fundamental physical constants. Some of the new physical relationships are shown in equations (14)-(26). For explanation of terms, refer to the definition of quantities below:

$$S = a_f^3/R_\infty = (q/D)^{1/2} = (i/J)^{1/2}, \tag{14}$$

$$\lambda = \alpha^2_y/2R_\infty = S/2\alpha_f, \tag{15}$$

$$\alpha_f = 1/2\theta = S/2\lambda = tf/2 = (SR_\infty)^{1/3}, \tag{16}$$

$$\theta = \lambda/S = \omega/f = T/t, \tag{17}$$

$$L = S = B/J = a_f^3 u/R_\infty, \tag{18}$$

$$C = S\varepsilon \tag{19}$$

$$Y = S/LC = S\alpha = (P_yZ)^{1/2} = (V_g\alpha)^{1/2} = R_\infty V_g/a_f^3, \tag{20}$$

$$h = q\Phi\theta = V_0 P_r/f = \Phi i/f, \tag{21}$$

$$m = q^2/S, \tag{22}$$

$$V_g = YS = A/LC, \tag{23}$$

$$\alpha = 1/LC, \tag{24}$$

$$d = DE/Vg = BH/V_g, \tag{25}$$

$$\eta = EH/Y = W/Q, \tag{26}$$

Note: One of the key elements of LSPR is that it shows visually how all physical quantities are interrelated and classified among themselves and how the same essence—the same reality—appears in different forms. Further, the different forms—physical quantities and constants—once decoded and classified, have a very orderly and logical structure. For example, it is well-known that when an area bounded by a curve is rotated about an axis, a volume is produced. Similarly, other physical quantities can be considered in this way and predicted from physical relationships. Listed below are a few examples from the left side of LSPR:

Volume (V_0)	= Angular area	($A \times S$),	(27)
Capacitance (C)	= Angular permittivity	($\varepsilon \times S$),	(28)
Self inductance (L)	= Angular permeability	($\mu \times S$),	(29)
Electric dipole moment (e)	= Angular electric flux	($q \times S$),	(30)
Magnetic dipole moment (b)	= Angular magnetic flux	($\Phi \times S$).	(31)

Please note that all physical quantities on the left side of LSPR are complementary to those on the right side of LSPR. Hence, to produce an angular quantity, when going from the left side of LSPR to the right side, instead of multiplication by S, we would divide the quantity by S. Listed below are a few examples from the right side of LSPR:

Left Side of LSPR

Voltage density (Y)	= Angular electric field strength (E/S),	(32)

Current density (J) = Angular magnetic field strength (H/S), (33)

Mass Potential Density (á) = Angular acceleration (y/S), (34)

Electric Field strength (E) = Angular electric potential (V/S), (35)

Magnetic field strength (H) = Angular magnetic potential (i/S), (36)

Mass field strength (acceleration) (Y) = Angular mass (gravitational), Potential (Vg/S). (37)

Please notice that as a rotational line produces an area, and a rotational area produces volume, in the same way the rotation of electric and magnetic potential produces electric and magnetic field strength (35), (36), and rotation of electric and magnetic field strength produces voltage and current density (32), (33). Because of space limitations, we are unable to describe the LSPR more in detail.

Summary: Views and appraisal. Using scale invariance as a fundamental concept in physics, we propose an approach for the integration of a scattered and immense body of fundamental physical phenomena into a more systematic order. The approach permits analysis of every corner of physics. Also, it enables us to predict new physical relationships and new constants.

The key to our approach is the Absolute principle—the scale invariant equilibrium frame of reference between all physical quantities and relationships. The Absolute frame of reference, we propose, is the ultimate foundation of nature, which includes within itself all physical quantities and serves as the ultimate principle out of which all physical quantities and relationships emerge (2)-(3). It is dimensionless (formless, timeless, massless, spaceless, etc.). Mathematically, the Absolute principle can be represented as «1», or on a logarithmic scale as «zero».

Just as each number, on a mathematical scale of numbers, has zero as its starting frame of reference, so also each physical quantity, on the physics scale of quantities, has the Absolute as its starting frame of reference. The physical quantities, we propose, are natural information units of measurement of the same essence—the same reality in different containers (quanta). Like chemical elements that can be arranged on a periodic table, these quantized information constituents of nature can be arranged in order of size on a logarithmic scale as demonstrated by the physics spectrum of quantities (Figure 1). The physics spectrum, through the logarithmic slide rule of physical relationships (LSPR), can then be used to study time, space, energy, and other quantities from various vantage points, from the smallest scale of quantum electrodynamics to the largest scale of cosmic voids, and to peer forward into the future for the derivation of new constants and laws of nature.

Acknowledgment. I am deeply indebted to Professor W. Fushchych for many valuable and enlightening discussions.

Definition of Quantities

A Cross-sectional area in meter²/radian²
B Magnetic flux density (magnetic induction) in weber-radian²/meter² or tesla
C Capacitance in coulomb/volt or farad
c Charge density in coulomb-radian³/meter³
D Electric flux (surface charge) density in coulombs-radian²/meter²
d Mass (flux) density in kilogram-radian³/meter³
E Electric field strength in volts-radian/meter
F (Electrical) force in (kilogram-meter-radian/second²) or newton
f Frequency (linear) (resonance frequency) in cycles/second or hertz
G Conductance in ampere/volt or siemen
g Electric conductivity in mhos-radian/meter
H Magnetic field strength (intensity) in ampere-radian/meter
h Planck's universal constant of action in joule-second/cycle
I Moment of (rotational) inertia in kilograms-meter²/radian²
i Magnetic potential (electric current) in (coulombs-radian/second) or ampere
J Current (magnetic potential) density in amperes-radian²/meter²
L Coefficient of self-inductance in (volt-second/amp-radian) or henry
M Mass flux density in kilograms-radian²/meter²
m Mass (flux) in kilogram
P Electric power in joule-radian/second or watt
P_r Pressure (stress, energy density) in kilogram-radian³/meter-second² or pascal
Q Volume flow rate in meter³/second-radian²
Q_m Mass flow rate in kilogram-radian/second
q Electric flux (charge) in ampere-second/radian or coulomb
R Resistance (capacitive or inductive reactance) in volt/ampere or ohm
R_∞ Rydberg constant in cycles³/meter-radian²
r Reluctance in ampere-turn/weber

S Length (radius of gyration, particle radius) in meter/radian

S_p Intensity of the wave (Pointing vector, illuminance, energy flux density) in watt-radian^2/meter^2

T Period of the harmonic motion in seconds/cycle

t Time (angular) (time constant) in seconds/radian

U Electromagnetic moment in ampers-meter^2/radian^2

V Electric potential (emf) in joules/coulomb or volt

v Velocity (linear) of motion in meters/second

V_0 Volume in meter^3/radian^3

V_g Mass (gravitational) potential, linear stopping power in joules/kilogram

W Energy (moment of force, work) in kilogram-meter^2/second^2 or joule

Y Voltage (electric potential) density in volt-radian^2/meter^2

y Mass field strength (acceleration) in meter-radian/second^2

Z Constant of universal attraction (gravitational constant) in m^3/kg-rad^2-sec^2

α Angular acceleration (mass potential density) in radian-second^2

α_f Fine structure constant in cycles/radian

ε (Electric) permittivity (of free space) in farads-radian/meter

η Coefficient of viscosity in pascal-second/radian or poise

θ Angle of rotation in radian/cycle

Ω Circular wave number in radian/meter

λ Wavelength (arc length, circumference) in meters/cycle (wave)

ω Angular velocity in radian-second (Magnetic) permeability (of free space) in henry-radian/meter

μ (Magnetic) permeability (of free space) in henry-radian/meter

1   One (Absolute frame of reference) dimensionless

Note: The unit radian (rad), for plane angle, has historically been designated as a supplementary unit. In 1980, the International Committee for Weights and Measures determined that angle and solid angle are to be regarded as dimensionless derived quantities. According to the committee, the unit radian and steradian are equivalent to the number one (1) and may be omitted in the expression for derived units. For completeness of presentation, we thought, because the angle of rotation (expressed in radian per cycle) is a physical quantity, which like other physical quantities, enters into physical relationships, it should be included here. Hence, all symbols for physical quantities, where applicable, include the radian and cycle terms.

1. *Bedrij O., Fushchych W.* On the electromagnetic structure of elementary particles masess // Dopovidi Ukrainian Academy of Sciences. Ser. A.—1991.—N 2.—P. 38-40.
2. *Bedrij O.* Fundamental constants in quantum electrodynamics // Dopovidi Ukrainian Academy of Sciences. Ser. A.—1993.—N 3.—P. 40-45.

Institute of Mathematics,
Academy of Sciences of Ukraine, Kiev

Submitted 20.10.92

МАСШТАБНАЯ ИНВАРИАНТНОСТЬ, ПРИНЦИП УНИФИКАЦИИ,
ПОРЯДОК И ПОСЛЕДОВАТЕЛЬНОСТЬ ФИЗИЧЕСКИХ ВЕЛИЧИН
И ФУНДАМЕНТАЛЬНЫЕ КОНСТАНТЫ

На основании принципа масштабной инвариантности физических соотношений (уравнений, формул) предложен феноменологический подход для введения понятия «порядок» между различными физическими величинами. Установлены новые соотношения между различными физическими величинами. В рамках предложенного метода вычислены фундаментальные константы в квантовой электродинамике.

МАСШТАБНА ІНВАРІАНТНІСТЬ, ПРИНЦИП УНІФІКАЦІЇ,
ПОРЯДОК І ПОСЛІДОВНІСТЬ ФІЗИЧНИХ ВЕЛИЧИН
І ФУНДАМЕНТАЛЬНІ КОНСТАНТИ

На основі принципу масштабної інваріантності фізичних співвідношень (рівнянь, формул) запропонований феноменологічний підхід для введення поняття «порядок» між різними фізичними величинами. Встановлені нові співвідношення між різними фізичними величинами. В рамках запропонованого методу обчислені фундаментальні константи в квантовій електродинаміці.

INDEX

'1'

V

vacuum, 12, 14, 27, 95
  characteristic impedance of, 50, 52, 59, 71
velocity, 17, 30, 32, 36, 44, 50, 63, 66-68, 73, 80, 83, 103
  of light, 36, 42, 96
  linear, 36, 38
von Klitzing constant, 17, 34, 44, 54, 61
von Neumann, John, 86

W

Waller, Ivar, 85
wave function, 28-29, 31-32, 35, 51
Weinberg, Steven, 14-16, 64, 66, 80, 92

Weyl, Hermann Klaus Hugo, 33, 36
Wheeler, John Archibald, 10-11, 14, 16, 23, 29-30, 46, 81, 90, 92
Wigner, Eugene, 31, 46
Witten, Edward, 11, 37

Z

Zermelo, Ernst Friedrich Ferdinand, 86

PERSONAL NOTES

'1'

PERSONAL NOTES

PERSONAL NOTES

PERSONAL NOTES

PERSONAL NOTES

PERSONAL NOTES

PERSONAL NOTES

PERSONAL NOTES

PERSONAL NOTES

'1'

PERSONAL NOTES

PERSONAL NOTES

'1'

PERSONAL NOTES

PERSONAL NOTES

PERSONAL NOTES

PERSONAL NOTES

ABOUT THE AUTHOR

Orest Bedrij is an author and interdisciplinary researcher into the foundation and mathematization of physics, at the Institute for Advanced Study of '**1**'. In 1962, at the age of twenty-nine, he was IBM's technical director at the California Institute of Technology Jet Propulsion Laboratory. He was responsible for the development and integration of the Space Flight Operations Facility computer complex that controlled the first soft landing on the moon. For the past forty-five years, Bedrij has been doing research into the unity of nature, the science of awareness, and into the physics and philosophy underlying ultimate reality and the laws of physics—i.e., searching the answers to our questions about why nature is the way it is and who we really are by way of physics and direct experience.

www.ingramcontent.com/pod-product-compliance
Lightning Source LLC
Chambersburg PA
CBHW022008170526
45157CB00003B/1195